I0056665

Unified Field Theory (Academic Edition)

Extraterrestrial Technology

Author: XiangQian Zhang
Editor: Lynn Lou Beran
Translator: Lynn Lou Beran

Published by Hope Grace Publishing
Alexandria, Virginia, USA

Published by Hope Grace Publishing 2025
HopeGracePublishing.com
Alexandria, Virginia, USA

This work is based on the original writings of **Zhang XiangQian**, translated, edited, and prepared for publication by **Lynn Lou Beran**. All efforts have been made to faithfully preserve the original author's ideas and intentions while making the content accessible to a broader audience.

ISBN: 978-1-966423-04-1
Library of Congress Control Number: 2024925087

First Edition: 2025

Table of Contents

Abstract

This book presents Zhang XiangQian's groundbreaking *Unified Field Theory*, which challenges mainstream physics and offers a revolutionary understanding of the universe's fundamental forces. Zhang proposes that space itself is in continuous motion, expanding outward from objects at the speed of light in a cylindrical spiral. His theory unifies the electric, magnetic, gravitational, and nuclear forces, dismissing the weak force as a fundamental interaction.

Central to Zhang's theory is the concept that rest energy arises from rest momentum, derived from the outward motion of space. By addressing these critical concepts, this book offers a profound reinterpretation of time, energy, and mass, with potential applications in technology, space exploration, and communication.

Zhang's insights were developed following an extraordinary encounter with an extraterrestrial civilization. This *Academic Edition* (2nd edition), translated and refined by Lynn Beran, provides readers with a deeper, clearer understanding of Zhang's theory and its far-reaching implications for science and the future of humanity.

Preface

The concept of the Unified Field Theory was first proposed by Albert Einstein, who spent over 40 years attempting to unify the electromagnetic and gravitational fields but without success. Today, four fundamental forces are recognized in nature: the weak force, electromagnetic force, gravitational force, and nuclear force. While the electric and magnetic forces have been unified under electromagnetism, our understanding of the nuclear force remains incomplete. Mainstream scientists also assert that the weak force has been unified with the electromagnetic force.

This text challenges the prevailing view by arguing that the electric and magnetic forces are distinct. Furthermore, it proposes that the weak force is not a fundamental force but rather a combined effect of electromagnetic and nuclear forces. The theory presented here offers a unified explanation for electric, magnetic, gravitational, and nuclear forces, encapsulated in a single mathematical equation that reveals their interrelationships.

Unified Field Theory addresses fundamental concepts in physics, including time, space, motion, force, the speed of light, velocity, mass, charge, energy, and momentum. The completion of this theory holds profound implications for humanity, but it also presents significant challenges.

This text focuses solely on the simplest and most fundamental scenario—point particles moving in a vacuum. It does not address the behavior of objects with defined shapes in a medium. In this context, point particles are treated as idealized entities with no consideration for shape or volume. Discussing the geometric dimensions of objects violates the assumptions of this theory and is therefore irrelevant.

Unified Field Theory attributes all properties of point particles to their motion through space or the motion of the space surrounding them. Thus, the theory does not concern itself with the internal structure of particles. It primarily examines the motion of the surrounding space and could therefore be called **Spatial Kinematics**.

At its core, Unified Field Theory posits that the physical world is an illusion and that all physical phenomena are merely human descriptions of reality. Understanding this concept is critical — without grasping it, one cannot fully comprehend the theory. Special attention should be paid to the **Vertical Principle**, a challenging yet essential concept within this text.

Introduction

The Unified Field Theory has long been one of the most elusive goals in modern physics, with many of the greatest minds — Albert Einstein among them — dedicating their careers to achieving this unification. The task of linking the fundamental forces of nature, including gravity, electromagnetism, the strong nuclear force, and the weak nuclear force, into a coherent framework has, until now, remained unsolved. In this work, Zhang XiangQian presents an innovative approach to this problem, drawn not from formal academic training but from unique, profound insights acquired through an extraordinary experience.

Zhang XiangQian's Unified Field Theory challenges mainstream understanding by proposing a new paradigm in which space itself is not static but in constant motion, influencing all matters and energy. His theory posits that space moves outward from an object at the speed of light, in a cylindrical spiral motion. This outward expansion of space provides the foundation for explaining the behavior of objects, forces, and fields throughout the universe. Central to this theory is the revelation that while current physics recognizes rest energy, it has failed to understand that the root cause of rest energy is rest momentum, resulting from the constant outward movement of space at the speed of light. This insight redefines the relationship between energy, mass, and motion in profound ways.

In Zhang's theory, the four fundamental forces are identified as the **electric force, magnetic force, gravitational force, and nuclear force**, diverging from the mainstream understanding, which includes the weak force. Zhang explains why the weak force should not be considered a fundamental force, asserting that it is a byproduct of electromagnetic and nuclear interactions, rather than an independent force. This reinterpretation suggests new pathways for understanding and harnessing the forces of nature, with potential applications in fields as diverse as advanced technology, neuralink, space travel, and communication.

This book introduces Zhang XiangQian's Grand Unified Equation, which offers a mathematical model aimed at uniting these four fundamental forces. It also explores the deeper implications of this theory, including the nature of information storage in space and the technology behind artificial field scanning systems. While challenging conventional wisdom, this work provides a new lens through which to view the universe and its underlying principles.

Zhang's personal journey is as extraordinary as his ideas. With only a middle school education and a background in farming and welding, his unique insight into the cosmos emerged from an alleged encounter with an extraterrestrial civilization. His theoretical developments have since gained attention for their potential to revolutionize physics, offering new solutions to problems that have baffled scientists for decades.

This book, now in its Academic Edition, builds upon the original text with additional contributions and clarifications to strengthen the argument for a Unified Field Theory. As Zhang's translator and editor, I have sought to ensure that these profound ideas are accessible to a wider audience, without compromising the precision and depth of the original concepts.

$P=m(C-V)$

《统一场论》

张祥前(1967-)

统一场论空间螺旋式运动模型示意图

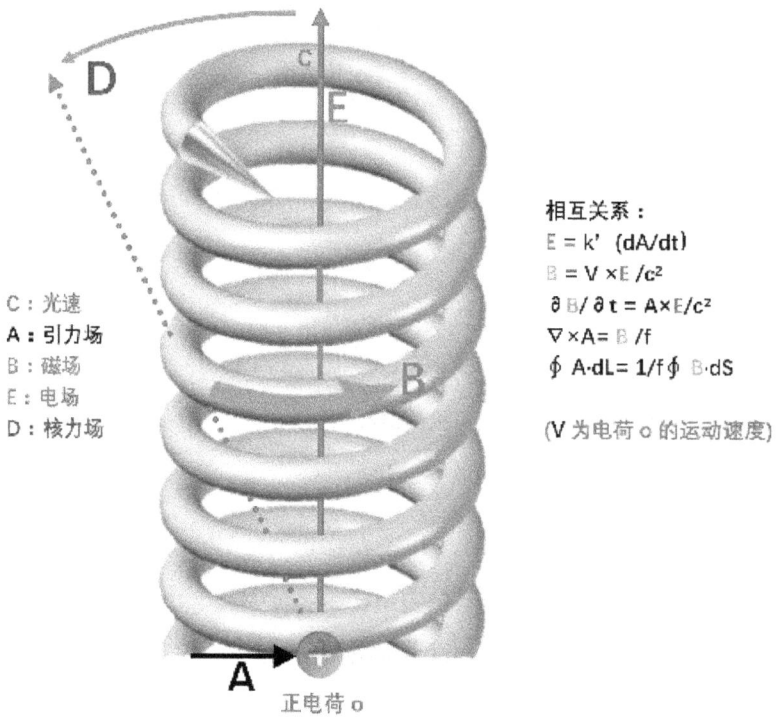

C：光速
A：引力场
B：磁场
E：电场
D：核力场

相互关系：
E = k' (dA/dt)
B = V ×E /c²
∂B/∂t = A×E/c²
∇×A= B /f
∮ A·dL= 1/f ∮ B·dS

(V 为电荷 o 的运动速度)

A

正电荷 o

Chapter 1. Composition of the Universe and Basic Principles of Unified Field Theory

The universe is composed of two fundamental elements: objects and the space that surrounds them. No third entity coexists with these two. All physical phenomena and concepts are merely human descriptions of the motion of objects in space and the movement of space itself. Without these descriptions by an observer, only objects and space truly exist in the universe—everything else ceases to exist and is merely the result of our interpretation of objects and space. The universe as we perceive it is illusory; the underlying reality consists solely of objects and space.

Objects and space do not consist of a more fundamental substance, nor can they transform into one another. Thus, the universe is inherently dualistic, not monistic.

The human brain constructs both the geometric and physical worlds through various interpretations of objects and space. When we describe the motion of objects and the motion of space, the physical world emerges. When we describe the size, quantity, direction, and structure of objects and space, the geometric world is born.

The physical world is primarily understood through human sensation, while the geometric world is understood through reason. Both the physical and geometric worlds are described from the perspective of an observer, and without this observer, neither would exist—only objects and space would remain.

The key distinction between the physical and geometric worlds lies in their descriptions. Physics primarily describes motion or the phenomena that result from motion, whereas geometry provides a more fundamental, straightforward processing of objects and space by the human brain. Physics represents a deeper and more complex processing, particularly focused on the description and analysis of motion.

Compared to physics, geometry encompasses a broader scope and is closer to the essence of the universe. As we know, mathematics includes both geometry and physics, and in this sense, physics can be viewed as the part of mathematics that describes motion.

Unified Field Theory does not attempt to answer why the universe consists of objects and space, or why they cannot transform into one another. It simply accepts these facts as the theoretical foundation for further reasoning.

The primary objective of Unified Field Theory is to explain the nature of fundamental physical concepts, such as time, displacement, mass, charge, gravitational fields, electric fields, magnetic fields, nuclear forces, energy, the speed of light, velocity, momentum, gravity, and motion, as well as the relationships between them.

Chapter 2. Definition of Matter

Matter is defined as that which exists objectively and independently of observers. In the universe, only objects and space exist in this true, independent sense. Thus, matter is composed solely of objects and space. Anything beyond this is merely a description created by human observers and does not exist independently of them.

For example, a tree or a river in front of us is "matter," while the growth of the tree or the flow of the river is an "event." In the universe, objects and space constitute "matter," while all other phenomena—such as time, displacement, mass, charge, fields, energy, the speed of light, velocity, momentum, force, temperature, or sound—are "events," properties described by observers due to the motion of matter relative to them.

This foundational principle denies that energy or time is part of matter and rejects the notion that fields are a distinct kind of matter. Instead, fields are either the result of the motion of material particles or the motion of space itself. According to Unified Field Theory, fields are simply effects caused by the changing motion of space.

Based on the principles of Unified Field Theory, concepts such as dark matter, dark energy, the God particle, gravitons, the ether, strings in string theory, and membranes are inferred to be non-existent—they are considered fabrications of human interpretation.

The space of the universe is infinite, as are the objects contained within it. Time, however, is merely a description of the human sensation caused by the motion of space and is a physical quantity defined by observers. As long as observers exist, time exists in the universe.

Moreover, the universe itself has no beginning or end. Its space and age are infinite. The Big Bang theory applies only to a specific region of the universe, and claiming that the entire universe originated from the Big Bang is incorrect.

Chapter 3. The Illusory Nature of the Physical World

The physical world is a product of our perception, created by the brain as it processes and describes the motion of objects and space. What we perceive—the physical world that we see and feel—is an illusion. It does not exist independently of us as observers. What truly exists is the underlying geometric world, composed solely of objects and space.

The geometric world is more closely aligned with the fundamental essence of the universe, whereas the physical world is primarily a description and interpretation of the geometric world created by the human brain. In this way, the physical world is a subjective construction, shaped by our perception of the motion and relationships between objects and space.

Chapter 4. How Physical Concepts Are Formed

It is meaningless to question how objects and space came into existence, as they are the most fundamental components of the universe. Objects and space cannot be composed of anything more basic. While objects can transform from one form to another, they do not arise or disappear without cause.

Objects and space inherently exist, just as the universe itself inherently exists. Therefore, questioning the origin or creation of the universe is also meaningless. We cannot define objects and space using something more fundamental because no such basis exists. However, we can define other physical concepts using objects and space as the foundation.

All physical phenomena and concepts ultimately arise from the motion of objects and space, as perceived by human observers. These concepts are the result of the brain processing and interpreting sensations of movement. Apart from objects and space, all other physical concepts—such as time, fields, mass, charge, the speed of light, force, momentum, and energy—are properties that emerge from the motion of objects through space or the motion of space surrounding objects, as observed by humans. These properties are fundamentally related to displacement and are formed through movement.

Time, fields, mass, charge, the speed of light, force, momentum, energy, and similar concepts can be considered functions of spatial displacement, and each can be expressed in terms of spatial displacement.

Sensory phenomena such as sound, color, force, and temperature are the result of objects moving through space and interacting with observers, triggering human sensations. These sensations are then processed and generalized by the brain to form physical concepts.

Fields and time, however, are somewhat unique. Fields represent the effects of the motion of space around objects, while time is the sensation we experience as the space around our bodies moves.

Chapter 5. Fundamental and Derived Physical Concepts

In physics, certain concepts are considered fundamental, while others are derived from these basic principles. For instance, time and displacement are fundamental concepts, while velocity is derived from the relationship between time and displacement.

Is there anything more fundamental than displacement and time?

Since the universe consists of two primary components—objects and space—these are the most basic physical concepts, serving as the foundational building blocks of the universe. They cannot be further defined, whereas other physical concepts can be derived from them.

Below is a diagram that illustrates the hierarchy of physical concepts, beginning with the most fundamental and building towards more complex derived concepts:

Objects (or particles), space → time, displacement, field → velocity, speed of light → mass, charge → momentum → force → energy, work → temperature, light, sound, color, etc.

This diagram demonstrates how higher-level, fundamental concepts like objects, space, time, and displacement give rise to lower-level, derived concepts such as velocity, mass, energy, and various sensory phenomena.

Chapter 6. Classification of Fundamental Physical Concepts

Fundamental physical quantities are divided into two primary categories: scalars and vectors. Scalars are quantities that can be represented solely by numbers, while vectors require both a numerical value and a direction to be fully described.

Scalars can further be classified into two types:

- **Positive and Negative Scalars**: Some scalars, such as charge, can take both positive and negative values. For instance, a positive charge is a positive scalar, while a negative charge is a negative scalar.

- **Purely Positive Scalars**: These are quantities that have no distinction between positive and negative values. Examples include physical quantities such as mass and energy, which are always positive.

Vectors, on the other hand, represent quantities that not only have a magnitude but also a specific direction. Examples of vectors include velocity, force, and displacement.

This classification helps distinguish between different types of fundamental physical concepts, aiding in their mathematical representation and understanding.

Chapter 7. How to Describe the Motion of Space Itself

Unified Field Theory asserts that space itself is in constant motion. While modern physics typically describes the motion of objects within space, the challenge lies in qualitatively and quantitatively describing the motion of space itself.

To address this, space is divided into many small sections, each referred to as a spatial geometric point, or simply a geometric point—also known as a space point. The path that a moving space point traverses is called a space line. By describing the motion of these space points, we can effectively describe the motion of space as a whole.

Mathematical methods commonly used in fluid dynamics and wave equations are equally applicable for describing the motion of space. In this context, space is regarded as a special medium, similar to a fluid. This allows us to use the mathematical tools developed for fluid mechanics to model and understand the movement of space.

Furthermore, Unified Field Theory recognizes that space exists independently and objectively. Its existence does not rely on the perception of an observer. Even in the absence of observers, space would continue to exist. However, time, as a construct defined by observers to measure motion, would not exist without people.

Chapter 8. Why Do Objects and Space in the Universe Need to Move?

Physics is our description of the geometric world, which consists of objects and space. Therefore, for any physical phenomenon, we can always find a corresponding geometric state. In physics, the motion states we describe are equivalent to the **perpendicular states** in geometry. If we do not describe them, the motion states are essentially the perpendicular states in geometry.

Note that part of this is reasoning, because every motion state must correspond to some geometric state. However, determining which geometric state corresponds to a motion state requires assumptions.

Unified Field Theory uses the **Perpendicular Principle** to explain why objects and space need to move. The **Perpendicular Principle** is described as follows:

Relative to us, the observers, any object in the universe can have up to **three mutually perpendicular straight lines** at any space point in the surrounding space. This is called the **three-dimensional perpendicular state of space**. Any space point in this perpendicular state must move relative to us, and the constantly changing direction of movement and the trajectory it follows can once again form a perpendicular state.

The above can be considered a qualitative description of the **Perpendicular Principle**. In the future, we also need to verify the quantitative description of the principle.

Motion with a constantly changing direction must be **curved motion**. Circular motion can have up to **two mutually perpendicular**

tangents. However, space is **three-dimensional**. At any point along the trajectory of motion, **three mutually perpendicular tangents** can always be drawn. Therefore, linear motion will inevitably be superimposed in the direction perpendicular to the plane of circular motion.

A reasonable view is that space points move in a **cylindrical spiral manner**, which is a combination of rotational motion and linear motion in the direction perpendicular to the plane of rotation.

Objects exist in space, and the positions of objects will change due to the influence of the movement of space itself. This explains why all objects in the universe need to move.

The belief that objects move due to **forces** is a superficial understanding. The underlying cause of all object motion in the universe is the **movement of space** itself. Conversely, we can use the movement of space to explain the nature of forces.

Objects can affect the space around them, and in turn, affect the objects that exist in that space. In this way, objects can interact through space without the need for any special medium to transmit interaction forces.

We must recognize that the motion of the space surrounding an object is caused by the object itself. Objects exist in space and can influence the surrounding space. The extent of this influence can be measured by the degree of motion in the surrounding space.

Objects exist in space and affect the surrounding space, causing the surrounding space to move. The movement of space inevitably affects the positions of other objects in that space, causing changes in their positions or creating a tendency for them to move.

All interactions between objects—**gravitational force, electric force, magnetic force,** and **nuclear force**—are essentially conducted through the movement of space itself. Objects transmit forces to each other through the changing space.

Space exists **objectively**, independent of observers. We can also regard space as a special kind of **medium**.

Which comes first, the motion of objects causing the motion of space, or the motion of space causing the motion of objects? They are **mutually causal**, with no primary or secondary relationship—objects and space are closely connected.

It is important to note that describing space motion has similarities and differences compared to describing the motion of ordinary objects. The space motion described in Unified Field Theory always refers to the space **surrounding objects**. Without objects, describing the motion of space alone is meaningless.

Describing motion requires determining the **initial moment of time** and the **initial position of space**. Pure space cannot determine the initial moment of time or the initial position of space. Determining these requires the joint determination of **objects and observers**.

The motion of space itself originates from objects and ends with objects. Without objects or observers, describing the motion of pure space is meaningless.

The **Perpendicular Principle** is one of the core secrets of the universe and is closely related to **spiral motion**. **Faraday's law** of electromagnetic induction in physics is also related to the **Perpendicular Principle**. The **vector cross product** and **curl** in mathematics are also related to the principle, though the proof is too complex to cover here.

Chapter 9. The Law of Spiral Motion

In the universe, everything—ranging from the smallest particles like **electrons, photons, and protons** to larger entities such as the **Earth, the Moon, the Sun,** and the **Milky Way**—moves in a **spiral pattern**. All freely existing particles in space, without exception, follow this motion. Even **space itself** moves in a **cylindrical spiral motion**.

The **Law of Spiral Motion** is one of the fundamental laws governing the universe. While it may appear that entities in the universe move in a **cyclic** manner, it is important to note that the universe is not a **closed system**.

The mathematical concept of the **vector cross product** is closely related to the law of spiral motion, though the proof of this connection is too complex to cover here.

Chapter 10. The Principle of Parallelism

In physics, the **parallel state** of two quantities corresponds to the **proportionality property** in mathematics. When two physical quantities can be represented by line segments and are parallel to each other, they must be in a **proportional relationship**.

This principle is closely related to the mathematical concept of the **vector dot product**, which expresses the relationship between two parallel vectors.

Chapter 11. Geometric Symmetry is Equivalent to Physical Conservation

In physics, **conservation laws** are directly related to **geometric symmetry**.

A **conserved physical quantity**, if it can be represented by a line segment, exhibits **linear symmetry** in geometric coordinates.

If the quantity can be represented by an area, it displays **plane symmetry** in geometric coordinates.

If it can be represented by a volume, it is **three-dimensionally symmetric** in geometric coordinates.

This fundamental connection between **symmetry** and **conservation** highlights the **geometric nature of physical laws** and the consistency of physical quantities across different dimensions.

Chapter 12. Continuity and Discontinuity of Space

The space that humans interact with and understand is generally considered **continuous**. Most mathematical systems used to model space, such as those found in classical physics, assume that space is continuous.

However, under certain conditions, space can exhibit **discontinuity**. For instance, when an object moves at the **speed of light** relative to an observer, the length of space in the direction of motion **contracts to zero**. This creates the appearance of discontinuity in the space where the object resides, as perceived by the observer.

This **discontinuity** of space is fundamental to phenomena such as **quantum entanglement** in quantum mechanics, where particles seem to interact instantaneously across vast distances, defying the limitations of continuous space.

The concept of space's **continuity and discontinuity** relates closely to both **relativity** and **quantum mechanics**. However, it is a broad and complex research area that will require years of investigation by many scholars, and thus will not be explored in detail here.

Chapter 13. The Description of Motion Cannot Be Separated from the Observer

Relativity posits that many physical concepts—such as **time, displacement, electric fields, magnetic fields, force,** and **mass**—are relative. Measurements of these quantities can vary depending on the observer's frame of reference. The term "relative" in this context means that these concepts depend on the **motion of the observer**.

Since concepts like **time, displacement, velocity, force, mass,** and **energy** arise from the motion of objects relative to an observer or the motion of the space surrounding the objects, describing motion without specifying an observer is meaningless. These physical concepts lose their significance if not tied to a particular observer's perspective.

At first glance, this viewpoint might seem **idealistic**. However, idealism claims that without observers, nothing exists—a notion that is also incorrect. The proper understanding is that all motion in the universe is **relative to the observer**. Without an observer, the universe would be like a **still frame** captured by a camera—static, but not non-existent.

In physics, when viewed geometrically, the state of motion is akin to a **perpendicular state**. These two phenomena are fundamentally the same, but observers perceive different results based on their angles of observation, whether from a physical or geometric perspective.

The state of motion is a product of our continuous affirmation and negation of an object's position in space. Some may argue that since the universe was in motion before humans existed, motion is independent of human observers. However, the phrase "**before humans existed**" is flawed.

If humans did not exist, how could there be a "before" or "after humans"? The very concept of "before" relies on human perception of time.

Terms like **"before," "after," "up," "down," "left," "right,"** and cardinal directions are all defined by human observers. Without us, these concepts cease to exist.

To describe motion in physics, the **observer**, the **object** (or particles), and **space** are all essential. Without any of these elements, the concept of motion loses its meaning.

The description of **time** is somewhat unique because the observer and the object of observation (our bodies) are the same. Human understanding of motion has evolved over time:

- **Newtonian Mechanics**: Newton's laws assume that an object's motion is described relative to a reference object considered to be at rest. Here, motion is defined by the distance an object travels in space over a period of time, and measurements of time and spatial length are independent of the observer's motion.

- **Relativity**: Einstein's theory builds upon Newtonian mechanics, emphasizing that different observers may measure different values for space and time. Relativity asserts that measurements of time and spatial length depend on the observer's velocity, with these effects becoming pronounced as speeds approach the **speed of light**.

Unified Field Theory posits that describing motion must always be **relative to a specific observer**. Without an observer, or without specifying which observer, describing motion becomes meaningless. In physics, both motion and rest are states that we, as humans, describe. Without us as observers, there is no state of motion or rest—only objects and space remain.

Whether objects and space are in a state of motion or rest cannot be determined without an observer. Thus, discussing motion or rest is meaningless without the context of observation. Furthermore, choosing a reference object to describe motion is sometimes unreliable.

Unified Field Theory also holds that **time is created by the observer's motion** through space, and is therefore linked to the observer's own movement. This means that **time is relative to the observer**, and different observers in relative motion may experience different durations for the same event.

Since **space itself is always in motion**, spatial displacement is also related to the observer's motion, and different observers may record different values for the same displacement. Like relativity, Unified Field Theory emphasizes that **your time and space,** and **my time and space** differ when we are in relative motion and should not be conflated.

Chapter 14. Why is Space Three-Dimensional?

We know that through any point in space, a maximum of **three mutually perpendicular lines** can be drawn. This defines the **three-dimensional nature of space**. But why exactly three lines, and not two or four?

The answer lies in the **motion of space**. If space moved in a straight line, it would create **one-dimensional space**. If space moved in a curved line, it would create **two-dimensional space**. However, space moves in a **cylindrical spiral motion**, which results in the formation of **three-dimensional space**.

The **three-dimensionality of space** is a direct consequence of this continuous cylindrical spiral motion. Since the three directions in space are of equal importance, with no single direction being unique or special, space moves equally in all three directions. This continuous motion leads space to naturally form a cylindrical spiral.

Figure 1: Cylindrical Spiral Motion of Space

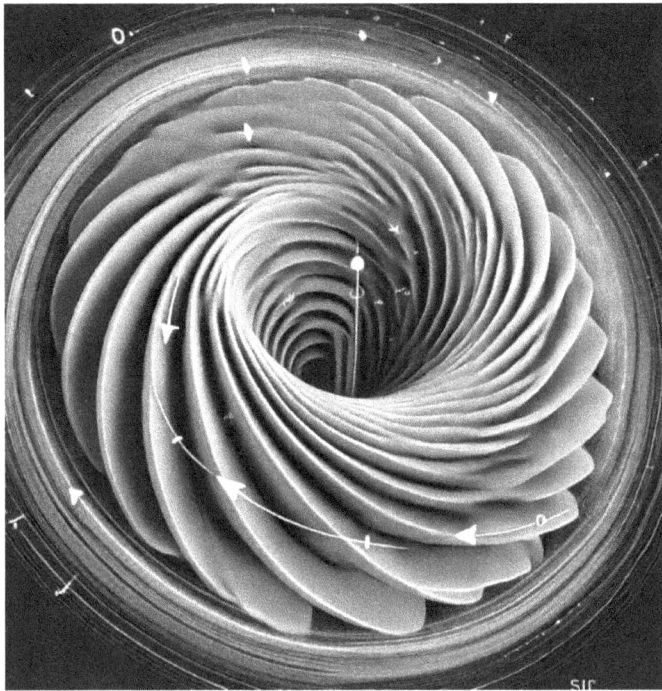

Alternatively, it can be stated that the **cylindrical spiral motion** of space creates three-dimensional space. These two concepts are causally related: the cylindrical spiral motion forms space, and space's three-dimensionality results from this motion.

The space we inhabit is referred to as **right-handed spiral space**. According to this, if the thumb of your right hand points in the direction of space's linear motion, the curl of your four fingers indicates the direction of space's **rotational motion**.

As for whether **left-handed spiral space** exists in the universe, it is theoretically repelled by the prevailing right-handed spiral space. Over billions of years, left-handed space would have been pushed to the farthest reaches of the universe. Even if it exists, it would be virtually undetectable to us.

When two **right-handed spiral spaces** (both appearing to rotate counterclockwise from our perspective) collide, the space where their rotations meet would contract, manifesting as **mutual attraction**. However, when left-handed spiral space encounters right-handed spiral space, they would **repel each other**.

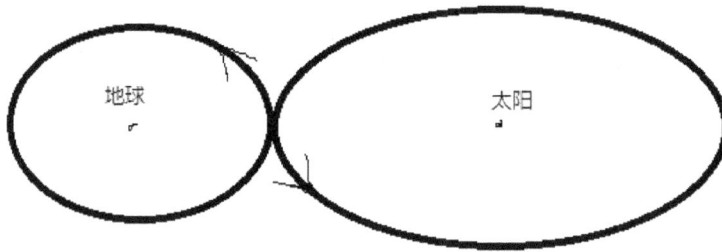

Interestingly, the space surrounding both **positive** and **negative charges** is also right-handed spiral. However, this aspect requires further theoretical and practical exploration. It is possible that in the future, humanity may be able to artificially create **left-handed spiral space**.

Chapter 15. Space Can Infinitely Store Information

Definition of Information: Information is the form of motion of matter, which consists of **objects and space**. The amount of information can be quantified by **possibilities**—the greater the number of possibilities, the greater the amount of information.

We can categorize the objects of our understanding into two types: **"events"** and **"things"**. Information belongs to the category of events.

While the amount of information stored or carried by any particle or object in the universe is **finite**, space in the universe has a unique property: any given space can store **all the information** from the **past, present,** and **future** of the entire universe. In other words, any finite region of space can contain an **infinite amount of information**.

This is because space is both **infinitely continuous** and **infinitely divisible**. The underlying reason for space's infinite capacity for storing information can also be understood logically:

The space surrounding an object expands at the **speed of light**, carrying with it all the information about the object into the surrounding space. Since space moving at the speed of light causes the length in the direction of motion to **contract to zero**, this space effectively becomes **two-dimensional**.

Thus, space moving at the speed of light can **instantaneously carry all information** about an object to any point in the universe, rather than spreading the information gradually at the speed of light, as is commonly assumed.

The universe consists solely of **two-dimensional** and **three-dimensional** space; there is no one-dimensional space or space with more than three dimensions. Since two-dimensional space has **zero volume**, it can maintain zero distance from any point in three-dimensional space. Therefore, information stored in two-dimensional space can permeate any point in three-dimensional space across the universe.

Conversely, it can also be said that **any point in three-dimensional space implicitly contains all the information from the past, present, and future** of the entire universe.

Why does it include **future information**?

Because **time is merely a perception** of us as observers. Without observers, time does not exist, and all information from billions of years ago and billions of years in the future can overlap at a single point in space.

Thus, the universe possesses not only **infinite time and space** but also **infinite information**.

This infinity of information can also be expressed in another way: The universe contains **infinite possibilities**. The repeated evolution of the universe is the expression of all these possibilities, occurring infinitely.

Information that arises in **three-dimensional space** can be stored in **two-dimensional curved space**—this can be rigorously proven through **Gauss's theorem** in field theory. Similarly, information that arises in **two-dimensional curved space** can be stored in **one-dimensional linear space**, which can be rigorously proven through **Stokes's theorem** in field theory.

It is important to note that the **generation of information** requires the participation of **material particles**. If material particles are excluded,

pure space cannot generate information, but it can transmit and store information. Furthermore, **information requires description by an observer**; without an observer, information does not exist.

Chapter 16. Basic Assumption of Unified Field Theory

Unified Field Theory begins with a fundamental assumption about the **motion of space** surrounding any object, including the body of the observer. When an object is at rest relative to an observer, the space around it moves outward in a **spiral manner**, centered on the object. This motion is a combination of **uniform rotational motion** and **uniform linear motion**, with the linear motion occurring **perpendicular to the plane of rotation**. This motion propagates at the **vector speed of light**, denoted as $\vec{c}$, where the magnitude of the vector (**c**) remains constant, but the direction of the vector speed of light can change.

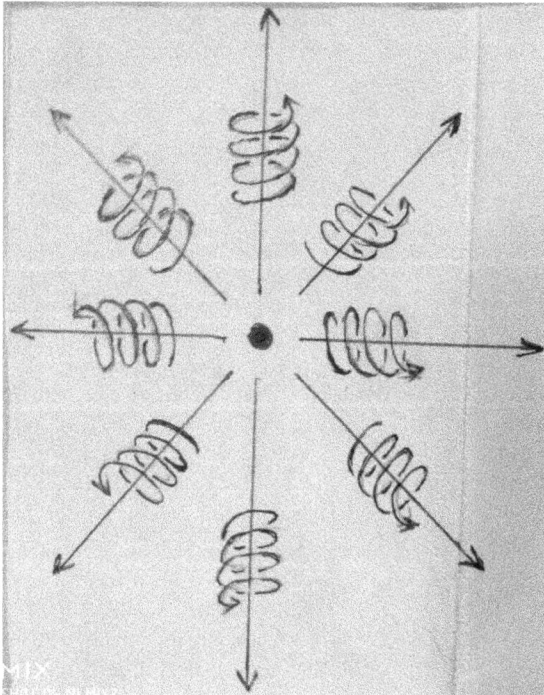

The surrounding space moves outward in a **cylindrical spiral motion**, carrying information and expanding from the object at the speed of light.

This assumption challenges the **Big Bang theory**, which posits a beginning and expansion of the universe. Unified Field Theory asserts that the universe has **no beginning or end** and has always existed in its current form.

One of the key pieces of evidence supporting the Big Bang theory is the observation that **space is expanding** relative to any observer. However, Unified Field Theory explains this expansion differently. The true reason for the expansion is that around every object in the universe, including the observer, space is expanding outward in a **cylindrical spiral at the speed of light**. Celestial bodies in space appear to move away from us as observers due to this outward spiral motion.

But why do bodies like the **Moon** and the **Sun** not move away from us at the speed of light?

This discrepancy is explained by the **initial state of motion** of the object or celestial body. For instance, the Earth has always been relatively stationary with respect to us as observers, and the Moon has been nearly stationary relative to us, particularly when compared to the speed of light. Only **distant celestial bodies**, which have little relation to our reference frame, move away from us at high velocities.

Chapter 17. The Physical Definition of Time

The fundamental principle of **Unified Field Theory** asserts that all physical concepts originate from our description of **motion** as observers. In the universe, there are two most basic forms of motion:

The **motion of objects** within space, and

The **motion of the space** surrounding objects.

The most fundamental physical concepts arise from these motions, as they create sensations for us as observers. When we analyze, describe, and generalize these sensations, we form the physical concepts that govern our understanding of the universe.

In our daily lives, we constantly perceive the **passage of time**. Time can be understood as a **sensation** that arises from either the motion of an object in space or the motion of the surrounding space itself. But what exactly is moving that creates this sensation of time?

Imagine sending a person in a spaceship to a region of space hundreds of billions of light-years away. After dropping the person off, the spaceship returns, leaving the individual alone in a vast space where all other celestial bodies are extremely distant. Despite this, the person would still experience the **sensation of time**.

So, what is moving in this situation to provide the sense of time? Since the person perceives their body as stationary, the only possible motion is that of the **space surrounding them**. Thus, time is our perception, as observers, of the motion of the space surrounding our bodies.

Building on the basic assumption of Unified Field Theory—that the space surrounding every object, including our bodies, spreads outward in a **cylindrical spiral manner** at the speed of light—we can now provide a **physical definition of time**:

Time is the perception of the motion of space, which moves outward from any object (including our bodies) in a cylindrical spiral manner at the **vector speed of light** $\vec{c}$. This motion of space gives us, as observers, the sensation of time.

Some might argue that **time existed in the universe before humans**, and thus, the view that time is merely a human sensation is incorrect. However, the phrase "before humans existed" is logically flawed. If humans did not exist, how could there be a "before humans"?

The logical error arises from excluding humans in the first step ("before humans existed") and then using humans in the second step to define "before." Once humans are excluded, they cannot be used as a reference to define temporal concepts.

Without observers, concepts such as **"before," "after," "up," "down,"** and even **time** itself lose meaning. **Time** is a physical concept born from our perception of the **motion of the space surrounding our bodies**.

Chapter 18. Spacetime Unification Equation

The physical definition of time in **Unified Field Theory** simultaneously defines the **speed of light**. In this framework, **time, space,** and the **speed of light** are deeply interconnected. The speed of light represents the unification of spacetime, meaning that the essence of time is our description of **space moving at the speed of light**.

In Unified Field Theory, the speed of light is extended to a **vector quantity**, denoted as $\vec{c}$, where the magnitude remains constant (**c**), but the direction can change depending on factors such as **time (t)**, the speed of the light source, and the motion of the observer.

The relationship is expressed as:

$$\vec{c} = c \times \vec{N}$$

Where:

- $\vec{c}$ is the **vector speed of light**,

- **c** is the **scalar speed of light**, and

- $\vec{N}$ is the **unit vector** indicating the direction.

The scalar speed of light **c** does not change with time, the motion of the observer, or the speed of the light source.

Based on the physical definition of time, it follows that:

- **Time is proportional** to the distance traveled by the space surrounding the observer at the speed of light.

- Using the concept of a **space point**, time is the sensation we experience as observers when numerous space points surrounding us spread outward in a cylindrical spiral manner at the vector speed of light c⃗.

For a space point **p**, starting at time zero from the observer's location and moving with vector speed c⃗, the time **t** experienced is proportional to the distance **R** traveled. This relationship leads to the derivation of the **spacetime unification equation**:

$$\vec{r}(t) = \vec{c}\,t = x\,\vec{\imath} + y\,\vec{\jmath} + z\,\vec{k}$$

Where $\vec{\imath}$, $\vec{\jmath}$, $\vec{k}$ are the unit vectors along the **x-axis, y-axis,** and **z-axis**, respectively. The scalar form of this equation is:

$$r^2 = c^2\,t^2 = x^2 + y^2 + z^2$$

These two equations can be considered **spacetime unification equations**, analogous to the spacetime relativity equations in Einstein's theory of relativity. They reflect the idea that **space and time share the same origin**. Time can, therefore, be represented by the **spatial displacement at the speed of light**.

It is important to note that not only **time**, but also other fundamental physical concepts such as **mass, charge, fields, momentum, force,** and **energy** are ultimately caused by and composed of **spatial displacement**. When tracing the essence of these concepts, we find that they can be reduced to and decomposed into spatial displacement.

This reflects the essence of physics: **Physics is the study of motion, and all motion is composed of spatial displacement.**

Chapter 19. Three-Dimensional Cylindrical Spiral Spacetime Equation

As discussed, all objects (or particles) in the universe, including space itself, move in a **cylindrical spiral pattern**. This spiral motion is a fundamental law of the universe. Unified Field Theory posits that space surrounding an object also moves in this cylindrical spiral manner.

To replace the four-dimensional spacetime equation of relativity, we establish the **three-dimensional cylindrical spiral spacetime equation** within Unified Field Theory.

Consider a particle at point **o** in a certain region of space, stationary relative to us as observers. Point **o** is taken as the origin of a **three-dimensional Cartesian coordinate system** with axes **x, y,** and **z**.

At time t'=0, we examine any space point **p** in the space surrounding the particle at point **o**, with its position represented by x_0 , y_0 , z_0. The spatial displacement vector from point **o** to point **p** (the position vector) is denoted as $\vec{r_0}$.

After a time period **t**, point **p** moves to a new position with coordinates **x, y,** and **z**. The position vector from point **o** to point **p** is now denoted as $\vec{r}$.

In **cylindrical spiral motion**, the motion can be decomposed into a **rotational motion vector** and a **linear motion vector**. Importantly, the position displacement (position vector) is a combination of these two vectors. According to the **Perpendicular Principle**, $\vec{r}$ changes with the spatial position **x, y, z**, and time **t**:

$$\vec{r}(t) = (x, y, z)$$

The specific relationship between **R(t)** and (x,y,z) is given by the **spacetime unification equation**:

$$\vec{r}(t) = \vec{r_0} + \vec{c}\, t = (x_0 + x)\, \vec{i} + (y_0 + y)\, \vec{j} + (z_0 + z)\, \vec{k}$$

This equation can sometimes be simplified as:

$$\vec{r}(t) = \vec{c}\, t = x\, \vec{i} + y\, \vec{j} + z\, \vec{k}$$

The scalar form is:

$$r^2 = c^2\, t^2 = x^2 + y^2 + z^2$$

r is the magnitude of the vector $\vec{r}$.

The above equation also appears in relativity, where it is considered the four-dimensional spacetime distance. However, in Unified Field Theory, **time** is described as the **motion of space** at the speed of light. When any one dimension in three-dimensional space moves at the speed of light, it is considered **time**.

Unified Field Theory emphasizes that **space** is fundamental, whereas time is an observer's description of space's motion. Without observers, **time** does not exist, but space remains. Relativity treats time as a fourth dimension alongside space, which is seen as a flaw in its interpretation.

Rotational and Linear Motion in the Three-Dimensional Spiral

If point **p** rotates in the **x-y** plane with an angular velocity ω and moves linearly along the **z-axis** at a constant speed **h**, with the projection length of **R** in the **x-y** plane being **R**, the motion is described by:

$$x = x_0 + R \cos(\omega t)$$

$$y = y_0 + R \sin(\omega t)$$

$$z = z_0 + h\,t$$

This can be written as a **vector equation**:

$$\vec{r} = \vec{r}_0 + \vec{c}\,t$$

$$= (x_0 + R \cos(\omega t))\,\vec{\imath} + (y_0 + R \sin(\omega t))\,\vec{\jmath} + (z_0 + h\,t)\,\vec{k}$$

Figure 2: Three-Dimensional Cylindrical Spiral Motion

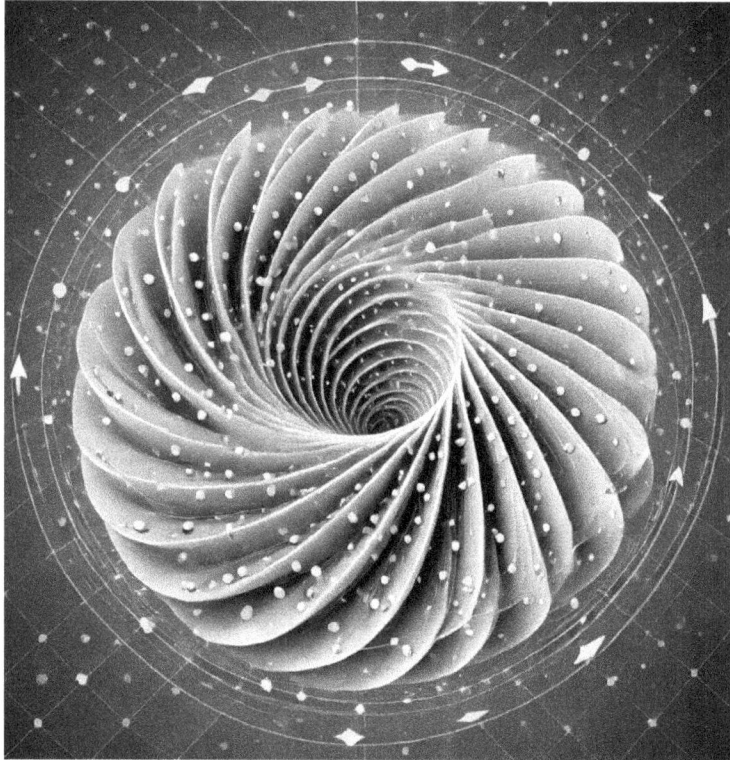

This is known as the **three-dimensional spiral spacetime equation**. It describes how objects move in space, combining rotational and linear motion.

In this equation, the relationship between **rotational motion** and **linear motion** is captured by the cross product:

$$x \times y = z$$

$$y \times x = -z$$

Here, **X** and **Y** represent rotational quantities, and **Z** represents the linear motion along the **z-axis**. These equations follow from the **Parallel Principle** and **Perpendicular Principle**.

Key Points to Note

i. There are many space points around point **o**, and the equation

$$\vec{r} = \vec{r}_0 + \vec{c}\, t$$

does not describe a single vector but rather a multitude of vectors radiating uniformly around point **o**.

ii. The **cylindrical spiral motion** of space is a combination of **linear** and **rotational motion**. Linear motion is a special case of cylindrical spiral motion where **R = 0**.

iii. The **field** is the result of space's cylindrical spiral motion. **Divergence** in field theory describes the linear motion component, while **curl** describes the rotational component.

iv. The spiral equation remains valid even when **x** and **y** approach zero (the motion is purely linear) or when **z** approaches zero.

v.   The **vector speed of light** is obtained by taking the derivative of the spiral motion equation with respect to time, considering both linear and rotational components.

vi.   **Spiral lines** emerge uniformly from the particle and trace paths based on the cylindrical spiral motion.

This three-dimensional cylindrical spiral spacetime equation serves as a fundamental law of Unified Field Theory, explaining not only the motion of celestial bodies and subatomic particles but also properties like mass and charge. The connection between **rotational** and **linear motion** is key to understanding this model of the universe.

Chapter 20. Understanding the Nature of the Speed of Light

1. The Nature of the Speed of Light

As physics continues to advance, the significance of the **speed of light** has gained increasing recognition, positioning it alongside other fundamental concepts such as **time, space, fields, mass, charge, momentum, force**, and **energy**. While people often associate the speed of light with the phenomenon of light itself, the speed of light reflects the fundamental laws of nature more profoundly than just the behavior of light.

In **Unified Field Theory**, the speed of light is extended to a **vector quantity**, expanding our understanding beyond its scalar form. Unified Field Theory offers deeper insights into the nature of the speed of light.

According to Unified Field Theory, the speed of light represents the **unification of space and time**. Space is fundamental, and its motion gives rise to **time**. Time, as perceived by observers, is simply our description of **space moving at the speed of light**.

The physical definition of time inherently binds **space, time**, and the **speed of light** together. By defining time through the motion of space, the speed of light is simultaneously defined. Space and time share the same origin, with the speed of light serving as the link between them.

Recognizing the speed of light as a **constant** demonstrates that space and time are, in essence, the **same entity**. When space extends, time extends correspondingly, and when space contracts, time contracts as well. This relationship is the essence of the unification of space and time.

The equation:

$$\vec{r}(t) = \vec{c}\,t = x\,\vec{\imath} + y\,\vec{\jmath} + z\,\vec{k}$$

is known as the **spacetime unification equation**. It illustrates the unified nature of space and time through the motion of space at the vector speed of light.

In the atomic scale, **electrons** exist in a confined space, moving at very high speeds with extremely short periods of motion. In contrast, **planets** in the solar system move in a vast space at much slower speeds and with longer periods of motion. This contrast in behavior is due to the **unification of space and time**.

At first glance, the **spacetime unification** in Unified Field Theory may appear contradictory to the **spacetime relativity** in Einstein's theory of relativity. However, in essence, they are consistent. The **spacetime unification equation** is fundamental, and from it, the **spacetime relativity equation** in relativity theory can be derived. The derivation process will be provided later in the text.

2. Explaining Relativity Effects Related to the Speed of Light

Why Is the Speed of Light Considered the Highest Speed in the Universe?

Relativity suggests that the speed of light is the maximum speed in the universe. This conclusion is primarily based on mathematical reasoning—exceeding the speed of light would cause certain physical quantities to become **imaginary numbers**, losing their physical meaning.

However, a simple logical explanation can also demonstrate that the speed of light is the upper limit. Consider an alien spaceship that is 10 meters long when stationary relative to us. As it accelerates, we observe its length contract, and when it reaches the speed of light, its length shortens to zero. If it were to travel faster than light, would its length become less than zero? Clearly not.

Similarly, relativity tells us that if a clock is placed inside the spaceship, and we hold another clock, both clocks would tick at the same rate when stationary. But as the spaceship moves relative to us, its clock slows down compared to ours. When the spaceship reaches the speed of light, its clock stops altogether.

For example, if an alien planet 50 light-years away sends a spaceship traveling at the speed of light toward Earth, it would take 50 years to reach us. However, the aliens onboard would experience the journey instantaneously—traveling an infinite distance in zero seconds. If faster-than-light travel were possible, would there be a speed faster than traveling an infinite distance in zero time? Clearly not.

These examples illustrate the well-known relativistic effects of **length contraction** and **time dilation**.

Are Length Contraction and Time Dilation Real Effects?

The idea that an object's length could shrink to zero, leaving it with zero volume, is hard for many to accept. Some argue that length contraction and time dilation are merely **observer effects**—perceptions caused by relative motion, not actual physical changes.

- One common belief is that these effects are **relative** to the observer outside the spaceship and that the spaceship itself does not change physically. Instead, light and electromagnetic waves reflecting off the spaceship give the illusion of contraction.

- Others argue that these effects occur **regardless of observation**. If there is relative motion, length contraction and time dilation are real phenomena.
- Some propose a **compromise**: that length contraction is an observer effect, while time dilation is a real effect.

Unified Field Theory takes a different stance, asserting that **both** length contraction and time dilation are real effects **and** observer effects. There is no absolute distinction between the two—they are **unified**.

Unified Field Theory's View on Observer Effects

In Unified Field Theory, the physical world we perceive is an illusion. Apart from **objects** and **space**, everything else in the universe—such as motion, time, mass, energy—is merely a description generated by our brains.

Space, according to Unified Field Theory, is created through **motion**. Space forms from positive charges radiating outward at the speed of light and converges at negative charges at the same speed. The **motion of space** requires description by observers. Without an observer, the motion and even the existence of time become meaningless.

This view challenges the traditional understanding of relativity, where space and time are considered fundamental. Unified Field Theory claims that **space** is fundamental, while **time** is an observer's description of space moving at the speed of light.

Real Effects Versus Observer Effects

In Unified Field Theory, concepts like **color**, **sound**, and **heat** are described as **observer effects**. For instance, heat is how our brain interprets the molecular motion of objects. Without the brain's interpretation, heat does not exist.

Similarly, the **state of motion** is also an observer's description. Both motion and rest are observer effects; without an observer, neither exists. The **only real entities** in the universe are **objects and space**—everything else is an observer effect.

Some might ask:

Some observer effects are consistent with real occurrences, while others are not. How do we distinguish between these two cases?

—There is no inconsistency.

What you see is what is happening. For something to really happen, there must be an observer to describe it; discussing a so-called real situation without an observer is meaningless.

Countless things happen in the universe every moment. When we discuss these things, we always relate them to a specific observer. Simply put, we say something relative to someone or something else.

If you don't specify to whom or to what something is relative, you often arrive at results that are ambiguous or misleading.

This is often why relativity is questioned or criticized — relativity is an incomplete theory. A complete theory should be unified field theory.

According to unified field theory, the existence of objects and space is an objective fact, independent of us as observers. Everything else is human description and subjective, belonging to observer effects.

In unified field theory, the effects of length contraction and time dilation can be applied concretely.

Unified field theory suggests that when an object moves at the speed of light, its length in the direction of motion shortens to zero, meaning it no longer occupies space. An object with zero volume could theoretically pass through a wall without damage to either the object or the wall.

Unified field theory can also explain space contraction due to motion using the perpendicular principle. Since the state of motion in physics is equivalent to the perpendicular state in geometry, when an object moves uniformly along the x-axis at any speed, the x-axis tilts. When the speed of light is reached, the rotation is 90 degrees, resulting in the projection of space along the x-axis in the direction of motion becoming zero.

In practical applications, unified field theory suggests that the mass and charge of an object are due to the space around the object dispersing at the speed of light, with the number of dispersing lines being proportional to the object's mass.

When an antigravity field generated by a varying electromagnetic field is applied to an object, it can reduce the number of light-speed-moving lines in the space around the object. When the number of light-speed-moving lines around the object decreases to zero, the mass becomes zero, and the object suddenly starts moving at the speed of light relative to us (this is the principle behind alien light-speed UFOs).

When the mass approaches zero, although the object may not move at the speed of light, it enters a quasi-excited state and can pass through walls without damaging either the wall or the object.

If length contraction and time dilation were purely observer effects, then the prediction by unified field theory that a rigid object could pass through a wall without damage would clearly be impossible.

Unified Field Theory and Mass at the Speed of Light

Relativity suggests that as an object approaches the speed of light, its **mass becomes infinite**. This is difficult for many to accept.

Unified Field Theory offers a different explanation: **mass** is a product of the motion of space around an object. As an object moves at nearly the speed of light, the space around it contracts, and the mass tends toward infinity due to relativistic effects. This explains why mass can be both infinite or zero, depending on the situation.

Some people think that when an object's mass reaches zero, the molecules within the object no longer interact with each other, causing the object to disperse like dust.

In this case, one observer might believe the object's mass is zero, while another observer might believe the mass is the same as usual.

This is different from the situation where, relative to any observer, the mass is zero.

Relativity suggests that if a spaceship moves at the speed of light relative to us, we observe that the spaceship's length in the direction of motion becomes zero, resulting in zero volume.

The observer inside the spaceship believes that there is no process between the start and end of the spaceship's motion, and the journey, no matter how far, is completed in an instant.

This is difficult for us to accept.

Unified field theory suggests that time is formed by the light-speed dispersion of space around the observer. When you move at the speed of light, you catch up with space, catch up with the light-speed motion of space, and thus catch up with time.

Therefore, from our perspective, you no longer have space, and your time has stopped, frozen.

This makes it easier to understand.

Relativity suggests that when an object moves at the speed of light, its mass becomes infinite, which is difficult for us to accept.

Unified field theory suggests that mass reflects the number of light-speed-moving lines of spatial displacement in the space surrounding the object within a certain solid angle.

When the object moves at nearly the speed of light, this solid angle becomes nearly zero due to the relativistic contraction of space, and the number of lines theoretically does not change with speed, causing the mass to tend toward infinity.

Since mass is a physical quantity observed by us and reflects the motion of the space around the object, and since the essence of mass is the effect of space motion, it becomes easier to understand why the mass of an object can be either infinite or zero.

The Speed of Light and Observer Effects

Speed is meaningful only in relation to an observer. In Unified Field Theory, **only the speed of light relative to the observer** is constant and the highest speed in the universe. Speeds and events unrelated to the observer are meaningless.

For example, if we rotate our bodies on Earth at a speed of one revolution per second, relative to an alien planet that is hundreds of billions of years old, the linear speed of the planet's rotation relative to us as observers would surely exceed the speed of light.

However, this superluminal speed has no causal connection to us as observers, so this superluminal speed is meaningless.

Another example, consider two spaceships traveling at 0.9 times the speed of light in opposite directions. An observer might calculate that their relative speed is 1.8 times the speed of light, but this speed is meaningless because it is not relative to the observer. Relative to us as observers, there is no superluminal speed.

Unified Field Theory also suggests that under certain conditions, the speed of light can be less than 300,000 kilometers per second. For instance, when the light source moves at a uniform speed relative to the observer, the speed of light along a perpendicular direction can be less than 300,000 kilometers per second.

3. Using the Physical Definition of Time to Explain the Invariance of the Speed of Light in Relativity

Relativity is founded on the **invariance of the speed of light**, but it does not explain **why** the speed of light is invariant. It simply accepts this invariance as a fact and uses it to modify Newtonian mechanics.

In relativity, the invariance of the speed of light means that:

- Whether the light source is stationary or moving at speed **v**, the speed of the emitted light **c** remains constant relative to us, the observers.

By understanding the **physical definition of time** in Unified Field Theory, we can clearly explain why the speed of light remains invariant.

The Physical Definition of Time

In Unified Field Theory, **time** is the sensation experienced by observers when the **space surrounding any object** (including our bodies) spreads outward at the speed of light **c**. This motion originates from the object at the center and light is stationary within this space, carried outward by the movement of space. The motion of this space gives us the sensation of time.

Thus, the amount of time **t** is proportional to the spatial displacement **r** of space moving at the speed of light **c**:

$$r = c \times t$$

The **speed of light** is then defined as:

$$c = \frac{r}{t}$$

From elementary mathematics, we know that a ratio is the **numerator divided by the denominator**. In this case, the **numerator** is the spatial displacement **r**, and the **denominator** is time **t**. These two quantities are fundamentally the same thing, merely given different names by us as observers.

For example, consider the analogy of **Zhang Fei** and **Zhang Yide**: these are two different names for the same person. If Zhang Fei gains 5 pounds, we immediately know that Zhang Yide also gains 5 pounds because they are the same individual.

Why the Speed of Light is Invariant

Since the **spatial displacement** (r) and **time** (t) are essentially the same entity, any change in the numerator (r) must result in a corresponding change in the denominator (t). Therefore, the **ratio** between them, **c**, remains constant.

When a light source moves relative to us at speed **v**, the numerator **r** changes, and the denominator **t** must change in sync, keeping the ratio constant. Whether the light source is moving at a constant speed or accelerating, this relationship holds, and the **speed of light remains invariant**.

For instance, if we observe that **Zhang Fei** gains weight, we know that **Zhang Yide** also gains weight, but the ratio between them stays the same because they are one and the same. Similarly, when the spatial displacement of light changes, time changes in the same way, ensuring that the speed of light remains unchanged.

Conclusion: Invariance and General Relativity

From this understanding, we can infer that whether the light source is stationary, moving at a constant speed, or accelerating relative to us, the speed of light remains constant. This confirms the correctness of **general relativity**, whose basic principle is that observers in relative accelerated motion will always measure the same speed of light.

Thus, the **physical definition of time** in Unified Field Theory provides a clear and logical explanation for the **invariance of the speed of light** in both special and general relativity.

Chapter 21. Explaining the Invariance of the Speed of Light in the Lorentz Transformation

1. Explanation of the Invariance of the Speed of Light in the Lorentz Transformation

Consider two right-angle inertial coordinate systems, s and s'. The spacetime coordinates of any event occurring at observation point p are represented as (x,y,z,t) in system s, and as (x',y',z',t') in system s'. System s' moves at a constant velocity v relative to system s along the positive x-axis, and at time t=t'=0, the origins of both systems coincide.

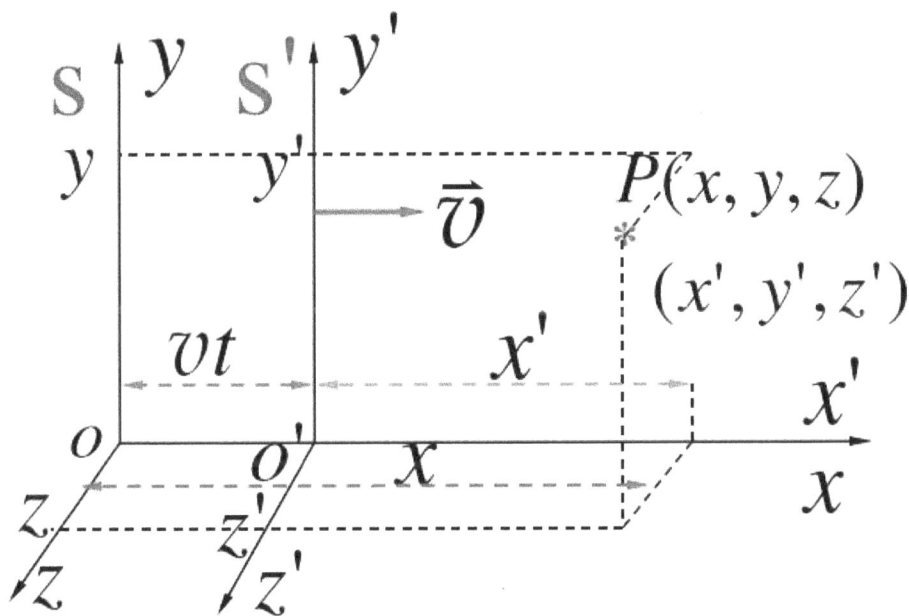

Imagine an event, such as an explosion, occurring at point p, which is stationary in system s'. In system s', the spacetime coordinates of the event are (x',y',z',t'), and in system s, they are (x,y,z,t).

In classical Galilean relativity, time is considered independent of velocity, meaning t=t'. However, special relativity shows that both time and spatial lengths depend on relative velocity. This dependence causes spatial lengths to contract as velocity increases.

For example, the spatial length along the x′-axis in system s must be shortened by a factor of 1/k, leading to the following relationship:

$$x' = k\,(x - v\,t) \qquad (1)$$

Similarly, in system s′, the spatial length along the xxx-axis must also be shortened:

$$x = k\,(x' + v\,t') \qquad (2)$$

To determine the value of kkk, we use the invariance of the speed of light.

2. Explaining Why the Speed of Light is Constant in a Single Reference Frame

One of the key insights from special relativity is that the speed of light remains constant within any given reference frame, regardless of the motion of the source or the observer. This constancy arises from the relationship between the motion of space and time surrounding the observer, expressed as:

Motion of space = c×Time

This equation illustrates that the speed of light c is a constant factor linking space and time. Therefore, within any single reference frame, the speed of light is invariant, regardless of the observer's velocity or the motion of other objects within that frame.

3. Explanation of the Invariance of the Speed of Light When the Motion of a Space Point is Perpendicular to Velocity v

Now consider the situation where the motion of a space point is **perpendicular** to the velocity v of the observed object. In this case, two reference frames, s and s', are aligned such that their xxx-axes coincide, and system s' moves at velocity v relative to system s along the x-axis.

At time t=t'=0, both system origins coincide. Afterward, system s' moves along the x-axis at a constant velocity v. Now, imagine an observer in system s' observing a space point p moving along the y'-axis at the speed of light ccc.

In system s, the distance op traveled by the space point in time t is longer than the distance o'p in system s'. Consequently, the time t observed in system s must also be longer than the time t' observed in system s'. This relationship is represented as:

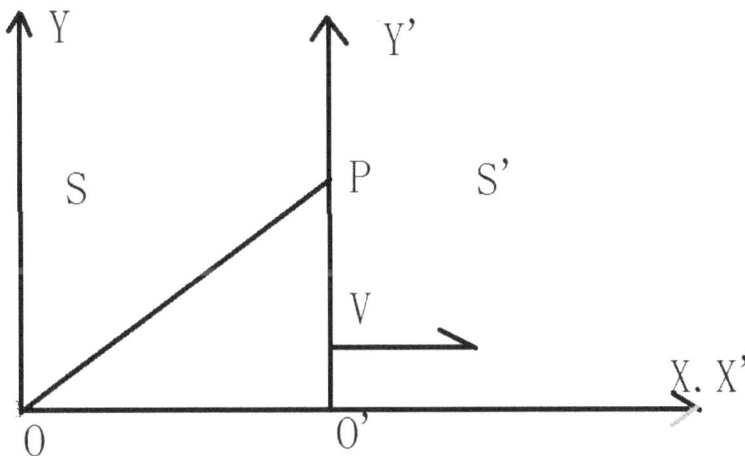

$$\frac{OP}{O'P} = \frac{t}{t'}$$

This relationship ensures that, even though the motion is perpendicular to the velocity of the moving frame, the speed of light remains invariant for both observers in systems s and s'.

4. The Relationship Between Light Source Velocity $\vec{v}$ and Vector Speed of Light $\vec{c}$

In Unified Field Theory, the **speed of light** is not merely a scalar quantity, but a **vector** $\vec{c}$, representing the direction as well as the magnitude of light's velocity. While traditional relativity treats the speed of light ccc as a scalar constant, the vector nature of light becomes important in certain relativistic contexts.

The magnitude of $\vec{c}$ remains constant at ccc, but its directional components along the x-, y-, and z-axes may vary depending on the velocity of the light source. The relationship between the velocity of the light source $\vec{v}$ and the vector speed of light $\vec{c}$ can be expressed using trigonometric relationships:

$$sin\beta = \frac{v}{c} \text{ or } cos\theta = \frac{v}{c}$$

Here, θ and β describe the angles between the velocity vector of the light source and the direction of light's vector speed. This vector-based approach provides a more nuanced understanding of light's behavior in moving frames.

5. Deriving the Invariance of Spacetime Intervals in Relativity

One of the most fundamental principles of relativity is the invariance of **spacetime intervals**. This means that the interval between two events in spacetime remains the same, regardless of the observer's motion. The spacetime interval Δs between two events is defined as:

$$\Delta s^2 = \Delta t^2 - \Delta x^2 - \Delta y - \Delta z$$

This equation is valid in both inertial systems s and s′. The invariance of Δs ensures that all observers, regardless of their motion, will agree on the "distance" between two events in spacetime. This invariance is critical for maintaining consistency in the laws of physics across different reference frames.

6. Correct Explanation of the Twin Paradox

The **twin paradox** is a thought experiment that demonstrates the concept of **time dilation** in special relativity. In this scenario, one twin travels at a high velocity while the other remains stationary. When the traveling twin returns to the stationary twin, they have aged less due to the effects of time dilation.

Breaking it Down:

1) What is Time Dilation?

- Time dilation refers to the phenomenon where time passes more slowly for an object in motion compared to one at rest. The faster the object moves, the more significant the effect. This is a direct consequence of Einstein's theory of special relativity.

2) Relating it to the Twin Paradox

- One twin stays on Earth, while the other twin travels through space at a speed close to the speed of light.

- According to special relativity, time for the traveling twin will **slow down** compared to the twin on Earth because the traveling twin is moving at a high velocity.

- Upon returning, the traveling twin will have aged less than the twin who remained on Earth. This is due to the high-speed motion experienced by the traveling twin, which affects the passage of time.

3) How Does Unified Field Theory Explain This?

- Unified Field Theory takes a slightly different approach. Instead of focusing solely on the concept of time slowing down, it relates the passage of time to the **motion of space** surrounding the observer.

- In this theory, the traveling twin experiences **less motion of space** during their journey because space is moving outward from them at a high speed.

- The twin who stays stationary on Earth, however, experiences more motion of space surrounding them.

- This difference in the motion of space results in **less time** passing for the traveling twin compared to the stationary twin.

4) What Happens at the Speed of Light?

- According to Unified Field Theory, when the speed of the traveling twin reaches the **speed of light**, the distance they experience in space becomes **zero**, and time essentially **freezes** for them.

- In this state, time no longer passes, and any motion through space is effectively halted. This means that from the perspective of the traveling twin, no time elapses, and no distance is traveled, even though they may appear to cover vast distances relative to the stationary twin on Earth.

5) *The Key Insight*

- The time dilation effect observed in the twin paradox is not just a consequence of high velocity, but of the **relative motion of space**. The faster the twin moves, the less space interacts with them, causing them to experience less time.

- As their speed approaches the speed of light, time slows to the point where it **freezes** and space **collapses** to a point. This explanation aligns with the **Lorentz transformation**, which mathematically shows how time and space are interconnected for objects moving at relativistic speeds.

6) *Conclusion*

- The twin paradox is resolved by recognizing that time is affected by both the **relative velocity** of the observer and the **motion of space** in Unified Field Theory.

- When the traveling twin approaches the speed of light, space distance shrinks to **zero**, and time freezes, offering a deeper understanding of the connection between the observer's motion and the behavior of space itself.

Chapter 22. General Definition of the Four Major Fields in the Universe

This chapter on the **General Definition of the Four Major Fields** in **Unified Field Theory** provides a foundational understanding of how fields are conceptualized mathematically and physically.

1. Mathematical Definition of a Field

- In mathematics, a field refers to a region of space where each point corresponds to a specific quantity. Fields can be:

  - **Scalar**: Each point in the field has a scalar value (e.g., temperature at a point).

  - **Vector**: Each point in the field has a vector quantity (e.g., the velocity of a particle at a point).

- A field is defined by a function at a given point in space, and mathematically, it is the **derivative of the displacement of space** with respect to position or time. In **Unified Field Theory**, this **displacement** refers to the **motion of space itself**, which moves outward from an object at the speed of light in a cylindrical helical pattern.

2. The Four Major Physical Fields

- **Unified Field Theory** identifies four primary fields in the universe:

  - **Gravitational**

  - **Electric**

  - **Magnetic**

o   **Nuclear** (strong force)

- These fields are not viewed in isolation but are the result of the **cylindrical helical motion of space** around particles.

- Importantly, the **weak force** is not considered a fundamental field in this theory but rather a **composite** of the electric, magnetic, and nuclear fields. Zhang's theory argues that the weak force emerges from the interactions of these fields rather than being independent, which differs from conventional physics.

3. Unified Definition of Physical Fields

- A **physical field** is defined as the **motion of space** surrounding a particle, relative to the observer. This motion is described as the **change** in the **position vector** from the particle to a point in space over time or across spatial coordinates.

- **Unified Field Theory** posits that **all physical phenomena** arise from either:

    o   The motion of particles in space.

    o   The motion of **space itself** around particles.

4. Fields as Moving Space

- The **unified definition** of the four major physical fields is as follows: Relative to the observer, the space Ψ surrounding a particle o at any point p in space, where the **displacement vector** $\vec{r}$ from o to p changes with spatial position (x, y, z) or time t, is called a **physical field** or **physical force field**.

- Mathematically, a field is the **derivative** of the displacement vector in the space around an object with respect to spatial position or time. This derivative measures the **motion of space** relative to the observer. In practical terms, the **four major fields** are defined by the **degree of motion** in the space around particles.

   Key Insight: **Fields are essentially moving space**. The **force** exerted by a field on an object is the result of space's motion, which changes (or has the tendency to change) the object's position in space.

- **Unified Field Theory** asserts that all four major physical fields are **vector fields** by nature. The difference between them is due to the observer's perception of the **cylindrical helical motion** of space from different angles and in different ways.

- The four essential elements for defining a field are:

   ○ **Space**

   ○ **Particle**

   ○ **Observer**

   ○ **Motion**

   Without these elements, the field would lose its meaning.

5. Three Forms of Field Descriptions

- Fields can be described in three primary forms:

   ○ **Distribution in 3D space**.

   ○ **Distribution on a 2D surface**.

- o **Distribution along a 1D curve**.

- These descriptions are linked through key **theorems** in field theory:

  - o **Gauss's Theorem**: Relates field distribution in space to surfaces using divergence.

  - o **Stokes's Theorem**: Relates field distribution on surfaces to curves using curl.

  - o **Gradient Theorem**: Describes scalar fields along curves.

  Examples:

The displacement of space around an object can be described as:

- A static three-dimensional region.

- A moving three-dimensional region.

- A static or moving curved surface.

- A static or moving curve.

Each description offers a different perspective on how the space surrounding a particle is displaced over time or in space.

6. Helical Motion and Field Dynamics

- **Unified Field Theory** posits that **fields emerge** from the **cylindrical helical motion of space**, which combines:

  - o **Linear motion** (captured by **divergence**).

  - o **Rotational motion** (captured by **curl**).

- This helical motion is the **foundation** for the dynamics of all four major fields. The **divergence** of a field measures the outward linear motion of space, while the **curl** captures the rotational component.

  Key Insight: The combination of linear and rotational motion in space defines the **structure and dynamics** of the four major physical fields. Understanding this helical motion is key to grasping how fields operate in **Unified Field Theory**.

7. Next Steps

This chapter has provided the basic framework for understanding the four major physical fields in **Unified Field Theory**. In the following chapters, we will delve deeper into the **precise definitions** and **behaviors** of each of these fields—gravitational, electric, magnetic, and nuclear—showing how they interact and how they differ from conventional field theory.

Chapter 23. Definition Equation for Gravitational Field and Mass

In Unified Field Theory, the mass m of an object at point o represents the number of spatial displacements r⃗ moving outward at the speed of light in a cylindrical helical pattern within the 4π solid angle surrounding point o.

The gravitational field A⃗ generated by point o represents the number of spatial displacements moving outward at the speed of light through the Gaussian surface s surrounding point o.

1. Definition Equation for the Gravitational Field

Consider a particle o, stationary relative to the observer. At time zero, any point p in the surrounding space departs from point o at the vector speed of light c⃗, moving in a cylindrical helical pattern along a certain direction. After a time t, it reaches the position where p is located at time t'.

Let point o be at the origin of the Cartesian coordinate system (x,y,z), with the vector r⃗ extending from o to p, following the previously introduced spacetime unification equation:

$$\vec{r}(t) = \vec{c}\,t = x\,\vec{i} + y\,\vec{j} + z\,\vec{k}$$

Thus, r⃗ is a function of spatial positions x,y,z, and time t, changing with x,y,z, and t, expressed as:

$\vec{r} = \vec{r}\,(x,y,z,t)$

The trajectory traced by point p in space is a cylindrical helical pattern, with the vector r⃗ tracing this helical path as one endpoint remains stationary at o while the other moves.

Using the scalar length r of $\vec{r}=\vec{c}t$ as the radius, we construct a Gaussian spherical surface s $= 4\pi r^2$ around particle o. The Gaussian surface must be continuous without any holes, but it need not be perfectly spherical.

Next, we divide the Gaussian surface s $= 4\pi r^2$ into small sections and select a small vector surface element ΔS at point p. The direction of $\vec{\Delta S}$ is represented by $\vec{N}$, with ΔS being the surface area magnitude. We observe that Δn displacement vectors $\vec{R}$, like p, pass perpendicularly through ΔS.

For convenience, we assume the radius of the Gaussian surface s equals the scalar length of $\vec{r}r$, allowing point p to lie on the Gaussian surface.

Thus, the gravitational field $\vec{A}$ generated by point o at point p is given by:

$$A = constant \times \frac{\Delta n}{\Delta S}$$

While simple, this equation does not fully capture the vector nature of the gravitational field or the spatial displacement $\vec{r}$ moving at the speed of light.

Vector Form of the Gravitational Field

To incorporate these details, we examine the situation at point p. The vector displacement $\vec{r}=\vec{c}t$ passes perpendicularly through $\vec{\Delta S}$. In general, $\vec{r}=\vec{c}t$ may form an angle θ with the normal direction $\vec{N}$ of $\vec{\Delta S}$.

When point o is stationary, the surrounding space motion is uniform, allowing $\vec{r}=\vec{c}t$ pass perpendicularly through $\vec{\Delta S}$, and the gravitational field A⃗ at point p can be written as:

$$\vec{A} = -\frac{G\,k\,\Delta n\,\vec{r}}{\Delta S\,r}$$

Where G is the gravitational constant, and k is a proportionality constant. The gravitational field A⃗ is in the opposite direction to the position vector r⃗ pointing from o to p.

Imagine n spatial displacement vectors r⃗ around point o, radiating outward in various directions. The meaning of n×r⃗ indicates that these displacement vectors are aligned and superimposed.

For Δn=1, the equation is simplified, but for integer n>1, it still holds physical significance.

$$\vec{A} = -\frac{G\,k\,\Delta n\,\vec{r}}{\Delta S\,r} = -\frac{G\,k\,\vec{r}}{\Delta S\,r}$$

The use of the unit vector r⃗ reflects that on the Gaussian surface s, we consider the direction and number of vectors r⃗, but not their length.

If r⃗ is not perpendicular to ΔS⃗, forming an angle θ with the normal N⃗, then with Δn=1, the equation can also be expressed using the dot product:

$$\vec{A} \cdot \Delta\vec{S} = -A\,\Delta S\,cos\theta = -G\,k\,\Delta n$$

Here, A is the magnitude of A⃗, determined by two factors: magnitude and directional cosine.

Finally, expressing ΔS in terms of the solid angle Ω and radius r, we get:

$$\Delta S = \Omega\,r^2$$

Thus, the gravitational field becomes:

$$\vec{A} = -\frac{G\,k\,\Delta n\,\vec{r}}{\Omega\,r^2\,r} = -\frac{G\,k\,\Delta n\,\vec{r}}{\Omega\,r^3}$$

2. Definition Equation for Mass

What is the essence of mass? What is the relationship between mass and the gravitational field?

Since the concept of mass originates from Newtonian mechanics, we can compare the geometric form of the gravitational field equation from Unified Field Theory:

$$\vec{A} = -\frac{G\,k\,\Delta n\,\vec{r}}{\Omega\,r^3}$$

with the gravitational field equation in Newtonian mechanics:

$$\vec{A} = -\frac{G\,m\,\vec{r}}{r^3}$$

From this comparison, we derive that the definition equation for the mass of an object at point o should be:

$$m = \frac{k\,\Delta n}{\Omega}$$

In differential form, this equation becomes:

$$m = \frac{k\,dn}{d\Omega}$$

In this equation, k is a constant, and since space can be infinitely divided, the differential dn is meaningful.

Taking the contour integral over the solid angle from 0 to 4π we get:

$$m = k \, \frac{\oiint dn}{\oiint d\Omega} = k \, \frac{n}{4\pi}$$

The physical meaning of this equation is that the mass m at point o represents the distribution of n spatial displacement vectors within the solid angle of 4π surrounding point o, where $\vec{r}=\vec{c}\,t$.

Thus, the differential definition equation for the geometric form of mass is:

$$m = \frac{k \, dn}{d\Omega}$$

In many cases, we set n=1, which simplifies the definition equation for mass to:

$$m = \frac{k}{\Omega}$$

Relationship with the Gravitational Field

Once we understand the essence of mass, we can explain the gravitational field equation in Newtonian mechanics:

$$\vec{A} = -\frac{G \, m \, \vec{r}}{r^3}$$

For example, consider Earth as point o, with an observer on the Earth. A satellite in orbit (point p) has a position vector $\vec{r}$ extending from point o to point p, with a magnitude r. The gravitational field generated by point o at point p, represented by:

$\vec{A} = -\frac{G \, m \, \vec{r}}{r^3}$ implies that on a Gaussian spherical surface s = $4\pi r^2$ with radius r, a small vector surface element $\Delta \vec{S}$ is cut out, with one vector

$\vec{r}$ passing through $\Delta S^{\rightarrow}$. The directions of $r^{\rightarrow}$ and $A^{\rightarrow}$ are opposite.

The reciprocal of the magnitude ΔS reflects the magnitude of the gravitational field, and the opposite direction of $\Delta S^{\rightarrow}$ indicates the direction of the gravitational field.

It's important to note that the gravitational field equation in Unified Field Theory reflects the situation at a specific moment in time.

Static Gravitational Field in Unified Field Theory

For the **static gravitational field** in Unified Field Theory, the equation:

$\vec{A} = -\dfrac{G \, k \, \Delta n \, \vec{r}}{\Omega \, r^3}$ can be further examined. When Δn and Ω are constants (i.e., when mass is constant), and only $\dfrac{\vec{r}}{r^3}$ is variable, the curl is zero:

$$\vec{\nabla} \times \vec{A} = \vec{0}$$

Similarly, when the mass $m = \dfrac{k \, \Delta n}{\Omega}$ is constant, and only $\dfrac{\vec{r}}{r^3}$ is variable, the divergence is also zero:

$$\vec{\nabla} \cdot \vec{A} = 0$$

However, when r approaches zero (i.e., when point p infinitely approaches point o), and point o can be considered an infinitesimally small sphere, the equation presents a 0/0 situation. Using the **Dirac delta function**, we obtain:

$\vec{\nabla} \cdot \vec{A} = 4\pi \, G \, u$ Where G is the gravitational constant, and $u = \dfrac{m}{\Delta x \, \Delta y \, \Delta z}$ represents the density at point o.

Thus, the curl and divergence of the gravitational field equation in Unified Field Theory are consistent with the corresponding equations in Newtonian mechanics.

3. Deriving the Relativistic Mass-Velocity Relationship from the Mass Definition Equation

In relativity, the mass-velocity relationship is commonly derived using the principles of momentum conservation and the relativistic velocity transformation formula. This relationship tells us that mass increases as the velocity of an object increases, eventually leading to the famous relativistic mass-energy equation. However, in this derivation, we will directly derive the mass-velocity relationship using the **mass definition equation** from Zhang XiangQian's theory.

The Setup

We begin by considering a particle with mass m', which is located at a point o in the stationary reference frame s'. The reference frame s' moves relative to another reference frame s at a constant velocity $\vec{v}$ (with magnitude v) along the positive x-axis. The x-axis of both s and s' coincide.

In the system s, the observer sees the mass of the particle as m. To find the relationship between m' and m, we will use the **mass definition equation**:

$$m \oint d\Omega = k \oint dn$$

where $d\Omega$ is the solid angle and dn is the number of geometric point displacements, which remains constant between the two reference frames.

The Solid Angle Ω

The solid angle Ω is defined as follows: On a spherical surface s centered at point o, with a radius r=1, we mark a small area ΔS. This area, with point o as the vertex, forms a cone h, and the area ΔS corresponds to the solid angle of the cone h.

The solid angle Ω of cone h is defined as the ratio of the base area ΔS of the cone to the square of the radius r of the sphere. As the area becomes infinitesimally small, we use:

$$d\Omega = dS/r^2$$

For a unit sphere with r=1, this simplifies to:

$$d\Omega = dS$$

Next, we extend this definition to the volume of the cone. For the same cone h, the volume ΔV of the cone is proportional to the solid angle. Therefore, the solid angle is also defined as:

$$d\Omega = \frac{dV}{r^3}$$

Again, for a unit sphere with r=1, this simplifies to:

$$d\Omega = dV$$

Mass Definition in Both Systems

Now, in the ss' system (where the particle is stationary), the mass m'
is given by:

$$m' = k \oint dn / \oint d\Omega'$$

Using the volume definition of the solid angle:

$$m' = k \oint dn / \oint dv'$$

where dv'=dx'dy'dz' is the differential volume in the stationary
system.

In the s system, where the particle is moving with velocity $\vec{v}$, the
mass is:

$$m = k \oint dn / \oint dv$$

Since the number n of geometric point displacements does not
change between the two reference frames, we only need to determine the
relationship between dv' and dv.

Lorentz Transformation

Using the Lorentz transformation for an observer in the s system,
moving at velocity $\vec{v}$, the spatial coordinates transform as:

$$x' = \frac{x - v\,t}{\sqrt{1 - \frac{v^2}{c^2}}}$$

$$y' = y$$

$$z' = z$$

Since the point of interest o is stationary in the s' system, , we focus on the relationship between differentials in the x-direction:

$$dx' = \frac{dx}{\sqrt{1 - \frac{v^2}{c^2}}}$$

while in the y and z directions, no transformation occurs:

$$dy' = dy$$

$$dz' = dz$$

Deriving the Mass-Velocity Relationship

Now, substituting this into the mass definitions in both systems, we have:

$$m' = k\oint dn/\oint dv' = k \oint dn/\oint dx'dy'dz'$$

$$m = k \oint dn/\oint dv = k\oint dn/\oint dx \, dy \, dz$$

from $\oint dx'dy'dz' = \oint dy \, dz \, dx/\sqrt{(1- v^2/c^2)}$

we can obtain:

$$m' = m \sqrt{1 - \frac{v^2}{c^2}}$$

This shows that the mass m' in the stationary system is related to the mass m in the moving system by the well-known **relativistic factor** $\frac{1}{\sqrt{1-\frac{v^2}{c^2}}}$, which is consistent with the theory of relativity. As the velocity v approaches the speed of light c, the mass increases significantly.

4. Lorentz Transformation of the Gravitational Field

Setup

Consider two inertial reference frames s′ and s, where s′ is moving with a constant velocity $\vec{v}$ (magnitude v) along the x-axis relative to s. In the frame s′, we have a **stationary, very thin rectangular panel** with mass that generates a **gravitational field $\vec{A}$** on its surface.

Gravitational Field Components in the x-Axis

We position the panel perpendicular to the x-axis.

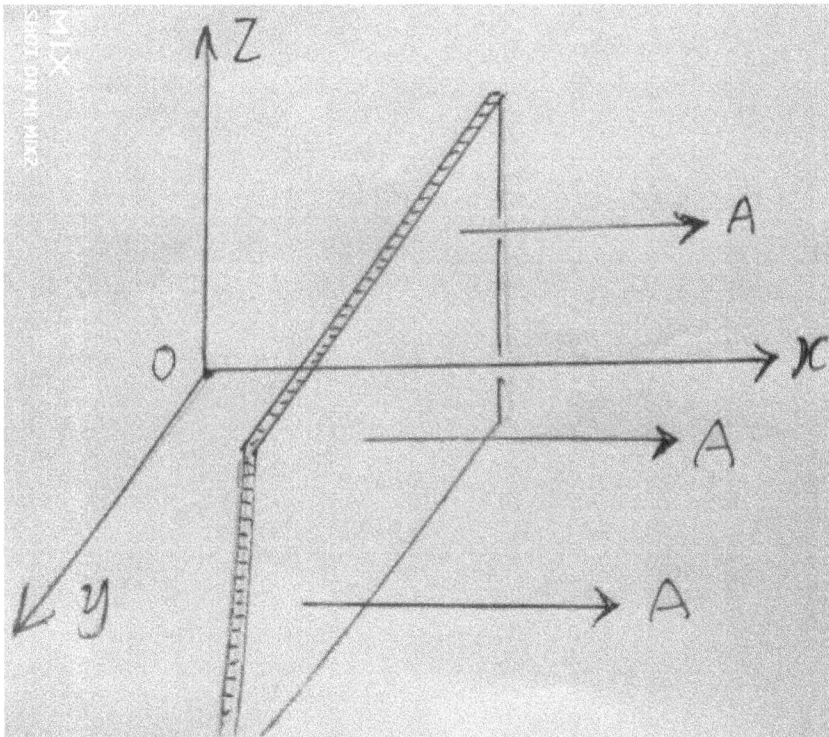

For an observer in frame s, the x-axis component of the gravitational field, Ax, appears unchanged. This is because the gravitational field strength is proportional to the number of spatial displacements passing through the surface, which is also proportional to the density. Since the area of the thin panel does not change, the number of displacements remains constant, and hence, the density remains constant.

However, due to the mass-velocity relationship in relativity, the mass of the panel increases by a relativistic factor

$$\gamma = 1 / \sqrt{1 - \frac{v^2}{c^2}}$$

. This increase in mass corresponds to changes in the spatial displacement vector's direction and the solid angle considered.

Thus, the x-component of the gravitational field in frame s becomes:

$$A_x = \frac{A'_x}{\sqrt{1 - \frac{v^2}{c^2}}}$$

where Ax' is the x-component of the gravitational field $\vec{A'}$ along the x'-axis in the s' frame.

Gravitational Field Components in the y and z Axes

Next, we position the panel parallel to the x-axis.

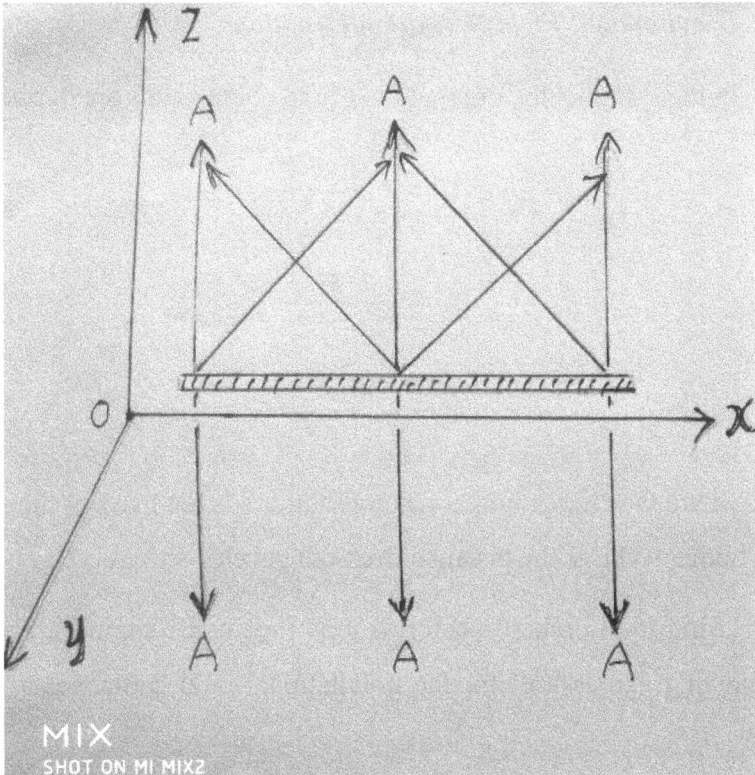

In this orientation, the panel contracts by the relativistic factor γ, and the mass increases by the same factor. Since the projections of the gravitational field lines on the x-axis cancel each other out, the x-component becomes zero. For the y and z components of the gravitational field, we obtain the transformations:

$$A_y = \frac{A'_y}{1 - \frac{v^2}{c^2}}$$

$$A_z = \frac{A'_z}{1 - \frac{v^2}{c^2}}$$

where Ay' and Az' are the y and z components of the gravitational field in the s' frame.

Gravitational Field Definition Equations

In the s' frame, the gravitational field components are defined by:

$$A'_x = -\frac{G\,m'\,x'}{r'^3}$$

$$A'_y = -\frac{G\,m'\,y'}{r'^3}$$

$$A'_z = -\frac{G\,m'\,z'}{r'^3}$$

where G is the gravitational constant, m' is the mass of the panel in the ss' frame, and r' is the distance from the panel.

Using the Lorentz transformations for the differentials, we derive the following expressions for the gravitational field components in the s frame:

$$A_x = -\frac{G\,m'\,x'}{\sqrt{1 - \frac{v^2}{c^2}}\,r'^3}$$

$$A_y = -\frac{G\,m'\,y'}{\left(1 - \frac{v^2}{c^2}\right)r'^3}$$

$$A_z = -\frac{G\,m'\,z'}{\left(1 - \frac{v^2}{c^2}\right)r'^3}$$

Full Gravitational Field in the s Frame

Using these results, we obtain:

$$A_x = -G \, m \, \gamma \, \frac{x - v \, t}{[\gamma^2 \, (x - v \, t)^2 + y^2 + z^2]^{\frac{3}{2}}}$$

$$A_y = -G \, m \, \gamma \, \frac{y}{[\gamma^2 \, (x - v \, t)^2 + y^2 + z^2]^{\frac{3}{2}}}$$

$$A_z = -G \, m \, \gamma \, \frac{z}{[\gamma^2 \, (x - v \, t)^2 + y^2 + z^2]^{\frac{3}{2}}}$$

Therefore, we express the full gravitational field in the s frame:

$$\vec{A} = -G \, m \, \gamma \, \frac{(x - v \, t) \, \vec{\imath} + y \, \vec{\jmath} + z \, \vec{k}}{[\gamma^2 \, (x - v \, t)^2 + y^2 + z^2]^{\frac{3}{2}}}$$

where $\gamma = 1 / \sqrt{1 - \frac{v^2}{c^2}}$.

Polar Coordinates Representation

Let θ be the angle between the radius vector $\vec{r}$ (magnitude $r = \sqrt{\gamma^2 \, (x - v \, t)^2 + y^2 + z^2}$) and the velocity $\vec{v}$. In polar coordinates, the gravitational field $\vec{A}$ is expressed as:

$$\vec{A} = - \frac{G \, m}{\gamma^2 \, r^2 \, (1 - \beta^2 \, sin^2 \theta)^{\frac{3}{2}}} \, \vec{e_r}$$

where $\beta = \frac{v}{c}$, and $\vec{e_r}$ is the unit vector of the radius vector $\vec{r}$.

This result mirrors the **relativistic transformation** form of the electric field, indicating that **Gauss's law** applies to both stationary and moving gravitational fields.

Divergence of the Gravitational Field

In the s′ frame, the divergence of the gravitational field is:

$$\vec{\nabla} \cdot \vec{A'} = \frac{\partial A'_x}{\partial x'} + \frac{\partial A'_y}{\partial y'} + \frac{\partial A'_z}{\partial z'} = \frac{G\, m'}{dV'}$$

where dV′=dx′dy′dz′ is the volume element in the s′ frame.

In the s frame, the divergence of the gravitational field becomes:

$$\vec{\nabla} \cdot \vec{A} = \frac{\partial A_x}{\partial x} + \frac{\partial A_y}{\partial y} + \frac{\partial A_z}{\partial z} = \frac{G\, m}{dV}$$

where dV=dxdydz is the volume element in the s frame.

Conclusion: Gauss's Theorem

Both forms of Gauss's theorem hold true, confirming that the theorem applies not only to stationary gravitational fields but also to gravitational fields in uniform motion. The differential form γdx=dx′ is derived from the Lorentz transformation x′=γ(x−vt) by taking the differential.

Chapter 24. Unified Field Theory Momentum Equation

1. Rest Momentum and the Nature of Motion

In **Unified Field Theory**, rest momentum plays a foundational role in explaining rest energy and motion momentum. **Rest momentum** is not a separate concept from motion momentum, and Zhang XiangQian's theory provides a framework that conserves momentum whether the object is stationary or in motion.

The Momentum of a Moving Object and the Quantity When Stationary Are Equal

Zhang XiangQian demonstrates that the **total momentum** of an object remains constant, whether it is at rest or in motion. While the **quantity** of momentum remains unchanged, the **direction** and **components** differ between the object's rest and motion states.

2. Momentum Equation in Motion

For an object in motion in the s frame, the momentum is described by:

$$\overrightarrow{p_{motion}} = m\,(\vec{c} - \vec{v})$$

Taking the dot product of both sides, we derive the magnitude of the momentum:

$$p_{motion}{}^2 = m^2\,(c^2 - 2\,\vec{c}\cdot\vec{v} + v^2)$$

Simplifying, we obtain:

$$p_{motion} = m \sqrt{c^2 - 2\,\vec{c} \cdot \vec{v} + v^2}$$

This equation shows that the total momentum in motion is influenced by both the object's velocity v and the speed of light C, but the **total magnitude** of the momentum remains constant as it transitions from rest to motion.

3. Equal Quantity of Rest and Motion Momentum

The **quantity of rest momentum** when the object is stationary is equal to the **quantity of motion momentum** when the object is moving, with the primary difference being the **direction** of the momentum vectors.

When the object is stationary, the **rest momentum** is given by:

$$\overrightarrow{p_{rest}} = m'\,\vec{c}$$

For the object in motion, the momentum is expressed as:

$$p_{motion}{}^2 = m^2\,(c^2 - 2\,\vec{c} \cdot \vec{v} + v^2)$$

$$p_{motion} = m \sqrt{c^2 - 2\,\vec{c} \cdot \vec{v} + v^2}$$

Thus, Zhang demonstrates that:

$$m'\,c = m \sqrt{c^2 - 2\,\vec{c} \cdot \vec{v} + v^2}$$

While the magnitude of the momentum remains constant, the **direction** shifts as the velocity v changes. This ensures that the total momentum is conserved across both rest and motion states.

4. Transformation Between Frames

Zhang's theory examines how the angle θ between the velocity vector v⃗ and the speed of light C⃗ approaches zero as the object's velocity approaches the speed of light. This ensures that **superluminal (faster-than-light)** speeds are avoided.

The relationship for the angle θ is derived using the **Lorentz transformation**:

$$\cos\theta = \frac{\cos\theta' + \dfrac{v}{c}}{1 + \dfrac{v}{c}\cos\theta'}$$

As v≈c, $\cos\theta$ approaches 1, and θ approaches zero, indicating that the object's motion becomes aligned with the speed of light as it approaches light speed.

5. Relativistic Mass-Velocity Relationship

From the equation:

$$m'\,c = m\,\sqrt{c^2 - v^2}$$

We derive the **relativistic mass-velocity equation**:

$$m = \frac{m'}{\sqrt{1 - \dfrac{v^2}{c^2}}}$$

This shows that the mass of the object increases as it moves, with this increase arising from the redistribution of the speed of light in the surrounding space. The total quantity of momentum remains conserved despite the change in mass and velocity.

6. Expansion of Momentum Conservation

Zhang extends the **conservation of momentum** across different reference frames, demonstrating that the total quantity of momentum measured for the same object remains unchanged regardless of the observer's relative motion. The principle here is that the **state of motion** can be observed but not altered by the observer.

7. Component Form Analysis

Breaking the momentum equation into components further clarifies the conservation of total momentum. The velocity u can be expressed as:

$$u = \sqrt{(c_x - v_x)^2 + \left(c_y - v_y\right)^2 + (c_z - v_z)^2}$$

Simplified, this becomes:

$$u = \sqrt{c^2 + v^2 - 2\,\vec{c}\cdot\vec{v}}$$

This component form highlights how the object's velocity components interact with the speed of light in different directions, maintaining the total momentum across all dimensions.

8. Energy and Momentum

By multiplying both sides of the equation $m' = m\sqrt{1 - \frac{v^2}{c^2}}$ by c^2, we arrive at the **relativistic energy equation**:

$$energy = m'\,c^2 = m\,c^2\sqrt{1 - \frac{v^2}{c^2}}$$

This shows that the increase in mass due to velocity corresponds to the energy required to maintain the object's motion, while the total momentum and energy remain conserved across reference frames.

9. Conclusion

Zhang XiangQian's **Unified Field Theory** redefines the relationship between rest and motion momentum, emphasizing the conservation of total momentum across reference frames. The magnitude of momentum remains the same whether the object is in motion or at rest, with the direction of the momentum vectors shifting as the object's velocity changes. By incorporating the **speed of light** and its interaction with space, Zhang offers a unified view of momentum, space, and energy, extending the classical and relativistic mass-velocity relationship.

Chapter 25. Unified Field Theory Dynamics Equation

1. General Assumption of Unified Field Theory

The central assumption of **Unified Field Theory** is that **space radiates outward** from any object in a **cylindrical spiral motion** at the speed of light. This outward motion is continuous and gives rise to both the object's **energy** and **momentum**.

- **Rest Momentum as the Root of Rest Energy**: Unlike current physics, which acknowledges rest energy without fully understanding its origins, UFT asserts that **rest momentum** is the cause of rest energy.

2. General Definition of Force

Force is defined as the change in the state of motion of an object relative to an observer over a specific range of space or time. This change can occur due to:

- The **movement of the object** through space.

- The **movement of space** itself around the object.

3. Types of Forces

- **Inertial Force**:

  o Arises from the derivative of the object's motion with respect to spatial position (solid angles).

  o Independent of the distance between the object and the observer.

  o Inertial forces can be generalized across classical, relativistic, and electromagnetic frameworks.

- **Interaction Force**:

  o Results from the derivative of the object's motion relative to spatial position, such as volume, surface area, or displacement vector.

  o Dependent on the distance between the force-exerting object, the object being acted upon, and the observer.

  o Interaction forces include **gravitational**, **electromagnetic**, and **nuclear** forces.

4. Unified Field Theory's Equation of Force

The **momentum** of a point, $p_{montion}$, around a particle o, is given as

$$p_{montion} = m\,(\vec{c} - \vec{v})$$

This equation reflects the assumption that space around the object is moving outward at light speed, and it is **independent of the distance** between points o and p.

- **Inertial force** is derived from the rate of change of momentum over time:

$$\vec{F} = \frac{d\vec{p}}{dt} = \vec{c}\,\frac{dm}{dt} - \vec{v}\,\frac{dm}{dt} + m\,\frac{d\vec{c}}{dt} - m\,\frac{d\vec{v}}{dt}$$

This equation incorporates the idea that **momentum changes** due to added mass or changes in velocity.

5. Interpretation of Forces in Unified Field Theory

The equation can be interpreted as follows:

- $(\vec{c} - \vec{v}) \frac{dm}{dt}$ is the **force due to added mass**, representing mass-change motion.

- $m \frac{d\vec{c} - d\vec{v}}{dt}$ is the acceleration force or the force due to changes in velocity.

These terms encompass four fundamental forces in Unified Field Theory:

- **Electric Field Force:** $\vec{c} \frac{dm}{dt}$

- **Magnetic Field Force:** $\vec{v} \frac{dm}{dt}$

- **Gravitational (Inertial) Force:** $m \frac{d\vec{v}}{dt}$

- **Nuclear Force:** $m \frac{d\vec{c}}{dt}$

6. Nuclear Force and Mass-Energy Equivalence

$\frac{d\vec{c}}{dt}$ is identified as the **nuclear force** due to its relationship to mass-energy equivalence:

$$E = m\,c^2$$

Work done by the nuclear force over a displacement $\vec{r}$ follows:

$$E = \int_0^r \vec{F} \cdot d\vec{r} = \vec{F} \cdot \vec{r} = m\,\vec{c} \cdot \frac{d\vec{r}}{dt} = m\,\vec{c} \cdot \vec{c} = m\,c^2$$

This aligns with energy released in nuclear reactions and atomic explosions, showing that the nuclear force is deeply connected to changes in space at the speed of light.

7. Mass-Change Motion and Discontinuities

The term $(c^{\rightarrow}-v^{\rightarrow})dm/dt$ also describes **mass-change motion**, which is **discontinuous** in nature. Specifically, when the **mass of an object is reduced to zero**, it reaches the speed of light instantaneously.

Quantum Mechanics Connection: This discontinuous nature of mass-change motion can explain the quantum behavior of energy radiation. A photon needs sufficient energy to reduce its mass to zero before it can travel at light speed.

8. Connection with Classical and Relativistic Mechanics

If space is assumed stationary, i.e., $c^{\rightarrow}=0$, the equation reduces to:

$$\vec{F} = -\vec{v}\,\frac{dm}{dt} - m\,\frac{d\vec{v}}{dt}$$

This is consistent with **classical mechanics** and **relativity**.

9. Conclusion

The **Unified Field Theory Dynamics** expands on classical force equations by incorporating the **light-speed motion of space** and the interaction between **mass** and the surrounding **energy fields**. The theory's dynamics equation unifies electric, magnetic, gravitational, and nuclear forces by describing their origin in the motion of space around an object.

- The **nuclear force** and **mass-change discontinuity** are key insights, explaining how forces emerge from the **outward motion of space**.

- This **unified framework** redefines how forces interact and offers a deeper understanding of the fundamental forces of nature.

Chapter 26. Explanation of Newton's Three Laws

Newtonian mechanics includes three fundamental laws and the law of universal gravitation. These laws are stated as follows:

i. **First Law (Law of Inertia)**: Every object (or particle) persists in its state of uniform motion in a straight line or in a state of rest unless acted upon by an external force.

ii. **Second Law (Law of Acceleration)**: When an object is subjected to a force, it accelerates in the direction of the force. The acceleration produced is directly proportional to the magnitude of the force and inversely proportional to the mass of the object.

iii. **Third Law (Action and Reaction)**: For every action, there is an equal and opposite reaction. When one object exerts a force on another, the second object exerts a force of equal magnitude and opposite direction on the first object.

Newton's mechanics, according to modern understanding, only hold true when considering a specific observer's frame of reference.

Newton defined the momentum $\vec{p}$ of an object with mass m and velocity $\vec{v}$ as $\vec{p}=m\vec{v}$. A closer analysis shows that the core of Newtonian mechanics is the concept of momentum, which originated from Newton's laws. We can now restate Newton's three laws using the concept of momentum:

Restated First Law (Inertia and Momentum):

Relative to a specific observer, any particle with mass m in space tends to maintain a definite momentum $m\vec{v}$. Here, $\vec{v}$ represents the velocity of the particle in a straight line, including the case where the velocity is zero (which means the momentum is also zero), corresponding to a state of rest.

Restated Second Law (Force and Momentum Change):

When a particle is subjected to an external force, it will cause a change in momentum. The rate of change of momentum $\vec{p}$ with respect to time t is equal to the external force applied:

$$\vec{F} = \frac{d\vec{p}}{dt} = \frac{d(m\,\vec{v})}{dt} = m\,\vec{a}$$

Restated Third Law (Conservation of Momentum):

The momentum of a particle is conserved. In an isolated system, when particles interact, the momentum gained by one particle is always equal to the momentum lost by another, ensuring that the total momentum remains constant.

In Newtonian mechanics, mass m is considered an invariant, while in relativity, mass can change. However, relativity retains many of Newtonian mechanics' concepts. The momentum formula in relativity has the same form as in Newtonian mechanics, with the key difference that mass m can be variable in relativity.

Unified Field Theory's Extension of Newton's Laws

Unified Field Theory, by revealing the nature of mass and space, offers a more thorough explanation of Newtonian mechanics. According to **Unified Field Theory**, Newton's three laws can be further understood as follows:

Restated First Law (Inertia and Light-Speed Dispersion):

Relative to our observer, the space surrounding any object itself radiates outward at a vector light speed $\vec{c}$. The number of light-speed moving spatial displacement lines n within the solid angle of 4π represents the mass m of the object:

$$m = k\,\frac{n}{4\pi}$$

Thus, when an object is at rest, it possesses a rest momentum $m\vec{c}$. To set this object in motion, an additional momentum (mass m multiplied by velocity $\vec{v}$) must be applied to alter $m\vec{c}$.

This extension of **inertia** reveals that even an object at rest has momentum due to the motion of space around it at light speed, giving rise to **rest momentum**.

Restated Second Law (Force as a Cause of Change in Motion):

Force is the cause that changes the state of motion of an object's surrounding space, which radiates at vector light speed $\vec{c}$ and moves with $\vec{v}$. In other words, force changes the total momentum, which includes both the object's motion and the motion of space around it. Thus, force is represented by the rate of change of momentum with respect to time:

$$\vec{F} = \frac{d(m\,(\vec{c} - \vec{v}))}{dt}$$

Force is defined as the change in the state of motion of the space surrounding the object. This motion is what ultimately alters the object's position or velocity in space.

Restated Third Law (Momentum Conservation in Unified Field Theory):

Momentum is the sum of:

- The motion of the object in space $m\vec{v}$, and

- The motion of the space surrounding the object itself $m\vec{c}$, expressed as $m(\vec{c}-\vec{v})$. This total momentum is a **conserved quantity**. The form of the momentum measured by different observers in relative motion may vary, but the total amount of momentum remains constant, independent of the observer's perspective.

This total momentum conservation law differs from classical mechanics by considering the contribution of **space's motion**. Thus, it provides a more complete view of the conservation principle across reference frames.

Conclusion

Unified Field Theory deepens our understanding of Newton's three laws by connecting them to the **motion of space itself**. This reinterpretation explains how **inertia, force,** and **momentum** are tied to the nature of space, offering a unified perspective that incorporates both classical and relativistic concepts into a more fundamental understanding of physics.

Chapter 27. Proof of the Equivalence of Inertial Mass and Gravitational Mass

1. Newton's Assumption

In classical physics, **Newton hypothesized** that **inertial mass** (the measure of resistance to acceleration) and **gravitational mass** (the source of gravitational force) are equivalent. However, Newton did not offer a theoretical explanation for this equivalence; he simply observed that the two quantities seemed to behave the same in all experiments.

2. Unified Field Theory Prove

In UFT, we seek to prove that **inertial mass** and **gravitational mass** are inherently the same by deriving their equivalence from the dynamics of space and mass interactions.

1) Newtonian Mechanics

- In Newton's framework, **inertial mass** reflects the difficulty of accelerating an object, while **gravitational mass** is the property of an object that allows it to exert gravitational force on another object.

- If we consider a point o with mass mmm and another point p with mass m', separated by a distance r, the gravitational force $\vec{F}$ experienced by point p due to the mass of point o is:

$$F = -\frac{G\,m\,m'}{r^2}$$

The gravitational force causes an **acceleration** $\vec{A}$ on point p, directed toward point o:

$$\vec{F} = -m' \vec{A}$$

- Newton, without offering an explanation, equated the inertial mass m′ in the equation $\vec{F} = -m' \vec{A}$ with the gravitational mass m′ in the equation $\vec{F} = -\frac{G\,m\,m'}{r^2}\,\vec{e_r}$,

- leading to:

$$\vec{A} = -\frac{G\,m}{r^2}\,\vec{e_r}$$

Here, r is the magnitude of $\vec{r}$, and $\vec{e_r}$ is the unit vector in the direction of $\vec{r}$. This equivalence between inertial mass and gravitational mass is widely accepted.

2) *Unified Field Theory*

- To prove that the acceleration $\vec{A}$ of point p towards point o equals the gravitational field generated by point o at point p, we proceed as follows: From the gravitational field equation:

$$\vec{A} = -\frac{G\,k\,n\,\vec{r}}{\Omega\,r^3}$$

- Here, n=1 is the number of spatial displacement vectors, and Ω is the **solid angle** surrounding point o, proportional to the area on a Gaussian sphere around point o.

- For simplicity, assuming the magnitude r of $\vec{r}$ remains constant while its direction changes, the gravitational field $\vec{A}$ becomes a function of the direction of $\vec{r}$ and the solid angle Ω. The solid angle Ω on a Gaussian sphere s= $4\pi r^2$ surrounding point o is proportional to $\vec{r} \cdot \vec{r} = c^2 t^2$. Thus, the gravitational field equation becomes:

$$\vec{A} = -\frac{G\,k\,\vec{r}}{c^2\,t^2\,r^3}$$

o   Since G, k, c, and r are constants, they can be combined into a single constant:

$$\vec{A} = -constant \times \frac{\vec{r}}{t^2}$$

o   Taking the second derivative of $\vec{r}$ with respect to time:

$$\vec{A} = -constant \times \frac{d^2\vec{r}}{dt^2}$$

o   Given that Newtonian mechanics is the earliest established system of mechanics, we can set the constant to 1, like how the proportionality constant in Newton's second law is set to 1:

$$\vec{A} = -\frac{d^2\vec{r}}{dt^2}$$

o   This completes the proof. This shows that the **acceleration** $\vec{A}$ due to gravitational force is identical to the change in the spatial displacement vector over time, which provides a theoretical basis for the equivalence of **inertial** and **gravitational mass**.

3. Inertial Force and Gravitational Force

- **Inertial Force**:
  - o   The **inertial force** is the force exerted by an object's **resistance to changes in motion**. It arises due to the object's mass and its tendency to resist acceleration.
- **Gravitational Force**:

- o The **gravitational force** originates from the **motion of space around an object** and influences nearby objects.
- **Unified Field Theory Explanation**:
  - o Both forces arise from the **same fundamental principle**: the movement of space at light speed and its interaction with mass.
  - o Since both inertial and gravitational forces depend on the same spatial dynamics, the **inertial mass** and **gravitational mass** are not just assumed to be equivalent—they are inherently **the same** according to UFT.

4. Conclusion

This chapter provides a **theoretical explanation** for one of Newton's long-standing assumptions, demonstrating that **inertial mass** and **gravitational mass** are equivalent under the principles of **Unified Field Theory**. By incorporating the **motion of space at light speed** and its interaction with mass, UFT shows that both types of mass are manifestations of the same underlying physical process. This proof strengthens the core of UFT by providing a unified framework for understanding the forces that govern motion and gravity.

Chapter 28. Explaining the Nature of Gravitational Force

One of the most perplexing questions about gravitational force is how it arises between any two objects in the universe and how it is transmitted between them. In fact, the essence of gravity is simple: it is the movement of space between mass points as perceived by observers.

Consider this analogy: when a car is driving towards you, the driver, feeling stationary, naturally perceives you as moving towards the car. If the car accelerates, the driver still perceives you as accelerating toward the car. Whether you are moving or the car is moving is irrelevant; the key is the **change in space between the car and you**. Similarly, gravity is the **movement of space between mass points**, as perceived by us observers. The relative motion of space between two mass points and their mutual movement are fundamentally the same thing.

Humanity has been misled by the term **force** in gravity, always pondering what this "force" is. The more we think about it, the more confused we become. Force is merely a **quality** we attribute to the relative motion between objects, just as we say a girl is beautiful or a knife is sharp. Force doesn't exist as a standalone entity but is a **description of relative motion**.

When two objects have relative accelerated motion or the tendency for such motion, we say they are under the influence of a force.

For example, if a person in China drops a ball, the ball accelerates towards the Earth. According to this view, one could also say the ball remains stationary while the Earth accelerates towards it. Similarly, if a ball were dropped simultaneously in Brazil, it would seemingly accelerate into the sky. This paradox results from the incorrect assumption that **space is stationary** and immobile, and that objects move through it like fish in a stationary ocean. In reality, **space itself is in constant motion** and closely linked to the movement of mass points.

1. The True Nature of Gravitational Fields

When we drop a stone on Earth, the stone falls freely due to the gravitational field. However, even **without the stone falling**, such as when the stone is hanging from a tree, the **space** it occupies still moves towards the Earth's center in the same manner. If space were visible, we would see it **falling towards the Earth**—this is the **true nature of the gravitational field**.

From the observer's perspective, the Earth causes the surrounding space to move uniformly, creating a uniformly distributed gravitational field. When another mass, such as a stone, enters this space, it disrupts the uniform motion of space between both the Earth and the stone. This **change in the uniform motion** within a solid angle is perceived as gravitational force.

Let's assign the stone as point p with mass m and the Earth as point o with mass m'. Based on Newton's laws and their redefinition through Unified Field Theory, the gravitational force $\vec{F}$ exerted by point o on point p is expressed as:

$$\vec{F} = m\,\vec{A}$$

where $\vec{A}$ is the gravitational field. The gravitational field produced by the Earth at point p (which is the **accelerated motion of space itself**) is equivalent to the acceleration experienced by point p. Thus:

$$\vec{A} = -\frac{G\,m'\,\vec{r}}{r^3}$$

where G is the gravitational constant, $\vec{r}$ is the vector from point o to point p, and r is the distance between the two points. From the relationship

$$\vec{F} = m\,\vec{A}$$

, we derive Newton's law of gravitation:

$$\vec{F} = -\frac{G\,m\,m'\,\vec{r}}{r^3}$$

This tells us that the **essence of gravitational force comes from relative motion**, aligning with the principle of Unified Field Theory: all physical phenomena arise from **motion**.

2. The Gravitational Field as Moving Space

The gravitational field around Earth, represented by

$$\vec{A} = -\frac{G\,m'\,\vec{r}}{r^3}$$

, measures the **degree of space's motion**. If another mass, such as point p, appears near Earth, the space around point p will adopt the same motion as the space surrounding Earth, altering Earth's gravitational field. The gravitational force $\vec{F}$ on point p is the degree of change in the gravitational field caused by the mass m of point p.

This change in the gravitational field does not refer to changes in time or space position, but to changes induced by the **mass** at point p. This is analogous to how multiplying one line segment by a perpendicular line segment transforms it into a rectangle.

3. Gravitational Field as a Conservative Field

In Newtonian mechanics, for a satellite orbiting Earth, the gravitational acceleration $\vec{A}$ directed towards Earth corresponds to the **gravitational field** at point p. If the satellite were removed, the space at point p would still accelerate towards Earth, representing the gravitational field.

In Unified Field Theory, the **gravitational field is the motion of space itself**. When the magnitude of the vector $\vec{r}$ remains constant and only its direction changes, the **curl of the gravitational field** becomes zero:

$$\oint \vec{A} \cdot d\vec{r} = 0$$

This indicates that the gravitational field is **conservative**.

Through cylindrical helical motion, the gravitational field reflects the first loop of space's spiral motion, pointing toward the center of acceleration. For instance, the Earth and Sun's mutual rotations cancel out some of the space between them, causing the space to contract and creating mutual attraction.

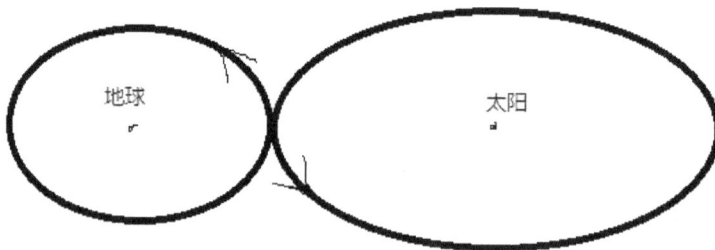

4. Conclusion

The **gravitational field** arises from the **motion of space** surrounding mass points, not from an intrinsic "force" but from the motion observed between objects. Unified Field Theory reinterprets gravity as a result of **space's spiral motion**, with the gravitational field behaving as a **wave** that propagates at the speed of light. The concept of gravity becomes a **dynamic relationship** between objects and space itself.

Chapter 29. Spacetime Wave Equation and Gravitational Field

Previously, we established that space surrounding an object radiates outward at the speed of light in a **cylindrical spiral motion**. The **vector displacement** of points in the space outside a mass changes with both its **spatial position** and **time**. This concept is crucial in understanding the wave-like behavior of gravitational fields.

1. Wave-Like Behavior of Space

Figure 3: Wave-Like Behavior of Gravitational Fields

When a physical quantity—such as the displacement of a space point outside a mass—varies with **spatial position** and **time**, it can be understood as undergoing a **wave-like process**.

- **Wave Motion vs. Cylindrical Spiral Motion**:

  o While **wave motion** typically involves the propagation of vibrations through a medium, **cylindrical spiral motion** refers to the displacement of points in space around a particle.

  o Despite these differences, in **space**, which is a special medium, **wave motion** and **spiral motion** can coexist.

- **Space Points**:

  o A single space point does not exhibit wave behavior, but a group of space points can show **wave-like characteristics**.

  o Unlike physical objects, **space points** are indistinguishable from one another, meaning that the **cylindrical spiral motion of space** contains **wave-like forms**.

2. Deriving the Spacetime Wave Equation

To better understand the nature of the **gravitational field**, we derive the **spacetime wave equation** from the previous **spacetime unification equation**:

$$\vec{r}(t) = \vec{c}\, t = x\,\vec{i} + y\,\vec{j} + z\,\vec{k}$$

where $\vec{r}$ represents the **displacement vector** of a space point p surrounding a mass o, and $\vec{c}$ is the **vector speed of light**.

3. Time Derivative of the Displacement Vector

Taking the derivative of $\vec{r}$ with respect to time t:

$$\frac{d\vec{r}}{dt} = \vec{c}$$

Squaring both sides:

$$\frac{d\vec{r}}{dt} \cdot \frac{d\vec{r}}{dt} = c^2 = \frac{dr}{dt} \cdot \frac{dr}{dt}$$

Here, c is the **scalar magnitude** of the speed of light, and r is the **scalar magnitude** of the displacement vector r.

4. Spacetime Wave Behavior of Another Space Point

Consider another space point p' moving around the central mass o. Let its displacement be represented by $\vec{L}$, which varies with time t. Given the relationship between $\vec{r}$ and t, we can deduce that $\vec{L}$ is also a function of $\vec{r}$.

Taking the second derivative of the displacement $\vec{L}$ of space point p' with respect to the scalar r of the spatial displacement vector $\vec{r}$, we get:

$$\frac{\partial^2 \vec{L}}{\partial r^2} = \frac{\partial^2 \vec{L}}{c^2\, \partial t^2}$$

This partial differential equation describes the **wave-like behavior** of space.

5. General Solution to the Spacetime Wave Equation

For the partial differential equation

$$\frac{\partial^2 \vec{L}}{\partial t^2} = c^2 \frac{\partial^2 \vec{L}}{\partial r^2}$$

The general solution to the spacetime wave equation is:

$$L(r,t) = f\left(t - \frac{r}{c}\right) + g\left(t + \frac{r}{c}\right)$$

where f and g are two independent functions.

The term $f\left(t - \frac{r}{c}\right)$ can be interpreted as a wave **propagating outward** from the point o in space. Conversely, $g\left(t + \frac{r}{c}\right)$ represents a wave **converging inward** toward point o.

6. Interpretation of Wave Behavior in Space

Traditionally in physics, the term $g\left(t + \frac{r}{c}\right)$, which represents a wave converging from infinity towards point o, is considered non-physical, as it seems to lack significance in ordinary media. However, in the context of **space**, which is a special medium, this wave behavior has physical meaning. This concept might also provide an explanation for the **origin of negative charge**, which will be discussed in more detail later.

7. Cylindrical Helical Wave Motion

The spacetime wave equation also describes **two forms of motion**:

1) **Radiation in straight lines** from the center point o in all directions.

2) **Convergence in straight lines** toward point o from all directions.

This motion can be viewed as the limiting case of a **spiral wave** with an amplitude approaching zero.

The partial differential equation:

$$\frac{\partial^2 \vec{L}}{\partial t^2} = c^2 \frac{\partial^2 \vec{L}}{\partial r^2}$$

has two specific solutions:

$$L = A \cos\left[\omega \left(t - \frac{r}{c}\right)\right] \text{ and } L = A \sin\left[\omega \left(t - \frac{r}{c}\right)\right]$$

These solutions describe **transverse waves** that propagate through space at the **speed of light**.

8. Cylindrical Helical Space-Time Wave Equation

Given the **continuity of motion**, the displacement components L_x and L_y along the x-axis and y-axis, respectively, should combine to form a **circular motion** in the plane perpendicular to the z-axis. Therefore, if Lx is taken as a cosine wave and Ly as a sine wave, we obtain the **cylindrical helical space-time wave equation**:

$$L_x = A \cos\left[\omega \left(t - \frac{r}{c}\right)\right]$$

$$L_y = A \sin\left[\omega \left(t - \frac{r}{c}\right)\right]$$

This describes the **helical motion of space** around a mass, with both transverse and spiral components.

9. Connection to the Gravitational Field

In **Unified Field Theory**, the **gravitational field** is the source of waves formed by **space vibrations**, while the **electromagnetic field** represents the propagation of these space vibrations at the speed of light. This framework unifies the concepts of **gravitational waves** and **electromagnetic waves** under a single theory, where both are forms of

space oscillations.

10. Conclusion

The **spacetime wave equation** derived from **Unified Field Theory** reveals that gravitational fields behave in a **wave-like manner**. These waves are not merely oscillations in a medium but reflect the dynamic motion of space itself, forming **cylindrical spiral waves** that propagate at the speed of light. This insight into the wave-like behavior of space provides a unified view of how **gravitational fields** interact with objects and propagate through spacetime.

Chapter 30. Definition of Charge and Electric Field Equations

1. Definition of Charge Equation

In **Unified Field Theory (UFT)**, both **charge** and **mass** are effects of the **motion of space** surrounding a particle at the speed of light, spreading outwards in a cylindrical spiral manner. Both originate from the same source—**space's light-speed spiral motion**.

Consider a particle o at rest relative to the observer. The **position vector** $\vec{r}$ points from point o to a surrounding space point p. We define a **Gaussian surface** $s=4\pi r^2$ around point o, where r is the magnitude of $\vec{r}$. Since point p is moving in a cylindrical spiral, its rotational motion traces out a **solid angle** Ω on the Gaussian surface s.

As previously stated, the mass m at point o can be expressed as:

$$m = k\,\frac{1}{\Omega}$$

Here, m represents the number of light-speed spatial displacement vectors $\vec{r}$ passing through the solid angle 4π surrounding point o.

In UFT, if the particle at point o carries a charge q, this charge represents the number of vectors $\vec{r}$ passing through the **unit solid angle per unit time**. Therefore, the rate of change of mass m with respect to time t corresponds to the electric charge. The definition of **electric charge** is given by:

$$q = k'\,\frac{dm}{dt} = -k'\,k\,\frac{1}{\Omega^2}\,\frac{d\Omega}{dt}$$

where k′ is a constant. This equation shows that the magnitude of charge is related to the **angular velocity** of the solid angle of rotational motion in the space surrounding a particle.

Since Ω is a **solid angle** and **4π** is one of its most significant values, this is the fundamental reason for the **quantization of charge**. The change in $\frac{d\Omega}{dt}$ represents a periodic change in the angle, implying that charge exhibits periodicity in time.

The essence of charge is closely related to the **rotational frequency of space**. This definition suggests that charge is the degree of rotational and outward helical motion of the space surrounding a particle at the speed of light. While this charge definition is part hypothesis and part inference, it aligns with the geometric structure of space in UFT.

2. Proof of the Invariance of Charge in Relativity

In **relativity**, charge is invariant regardless of the motion of the object. While this invariance is stated in the theory, it is not explicitly proven. Using the charge definition equation from UFT, we can prove this.

When a particle at point o is stationary relative to the observer and carries charge q, the relationship between charge and mass is:

$$q = k' \frac{dm}{dt}$$

When point o moves with velocity v relative to the observer, both the mass m and time t increase by a **relativistic factor** $\sqrt{1 - \frac{v^2}{c^2}}$. However, the charge q remains unchanged. This proves that **charge is invariant under relativistic transformations**, confirming that charge does not vary with the speed of the moving particle.

3. Considerations Regarding the Definition of Charge

In the definition of charge q, the term $\frac{dm}{dt}$ suggests that the charge of a particle is proportional to the **rate of change of its mass**. This might seem inconsistent with observation, as we do not typically see significant changes in the mass of charged particles.

One possible explanation is that the **mass change of charged particles** is periodic rather than continuous, with a frequency so high that the changes are imperceptible—much like **alternating current**. This periodicity may be connected to the **matter wave** in quantum mechanics, where particles are characterized by both **wavelength** and **frequency**.

4. Geometric Definition Equation of the Electric Field

Consider a particle o at rest, carrying charge q, which generates an **electric field** $\vec{E}$ at a point p in the surrounding space. The **position vector** from o to p is $\vec{r}$, and the magnitude of $\vec{r}$ is r.

According to **Coulomb's law**, the electric field is defined by:

$$\vec{E} = \frac{q\,\vec{r}}{4\pi\,\varepsilon_0\,r^3}$$

where $4\pi\,\varepsilon_0$ is a constant. Substituting the **charge definition equation**:

$$q = k'\,\frac{dm}{dt} = -k'\,k\,\frac{1}{\Omega^2}\,\frac{d\Omega}{dt}$$

into the equation for the electric field gives the **geometric definition of the electrostatic field**:

$$\vec{E} = -\frac{k'\,k}{4\pi\,\varepsilon_0}\,\frac{1}{\Omega^2}\,\frac{d\Omega}{dt}\,\frac{\vec{r}}{r^3}$$

This equation shows that the electric field represents the **density of space displacement** $\vec{r}$ passing through the Gaussian surface s per unit time. When the direction of the electric field aligns with the space displacement, it represents a **positive electric field**; when opposite, it represents a **negative electric field**.

5. Explanation of Coulomb's Law

Coulomb's law states that the **force** $\vec{F}$ between two stationary point charges q and q' in a vacuum is directly proportional to the product of their charges and inversely proportional to the square of the distance r between them:

$$\vec{F} = \frac{k\,q\,q'}{r^2}\,\vec{e_r} = \frac{q\,q'\,\vec{r}}{4\pi\,\varepsilon_0\,r^3}$$

where $\vec{e_r}$ is the unit vector along $\vec{r}$. The electric field $\vec{E}$ produced by charge q at the location of q' is:

$$\vec{E} = -\frac{k'\,k}{4\pi\,\varepsilon_0}\,\frac{1}{\Omega^2}\,\frac{d\Omega}{dt}\,\frac{\vec{r}}{r^3}$$

The presence of q' at point p near q changes the electric field $\vec{E}$ produced by q. This change in the field represents the **force** exerted by q on q', consistent with Coulomb's law.

6. Model of Positive and Negative Charges

In **Unified Field Theory (UFT)**, a particle carries an electric charge because the space surrounding the particle undergoes **cylindrical helical motion**. This motion can be decomposed into two components: **rotational motion** and **linear motion** perpendicular to the plane of rotation. We know that cylindrical helical motion is the result of the **superposition of rotational motion** and **straight-line motion** perpendicular to the plane of rotation, which can be explained using the **right-hand rule**.

A particle carrying a **positive charge** generates a **positive electric field** around it. The **linear motion** part of the space radiates outward from the particle's center in all directions, while the **rotational part** of this motion rotates counterclockwise, following the **right-hand rule**, creating the observed positive electric field. The **radial velocity** (a combination of the rotational speed and the linear motion speed) represents the speed of light, extending outward from the positive charge toward a point at infinity in space.

A **negative charge**, on the other hand, generates a **negative electric field**. In this case, the **straight-line motion** of the surrounding space converges inward from an infinite distance, while the **rotational component** still follows the **right-hand rule**, rotating counterclockwise. The **radial velocity** extends from infinite space toward the negative charge, representing the speed of light.

This **cylindrical helical motion** around charged particles explains why charges exhibit positive or negative behavior. For a **positive point charge**, drawing rays from the charge to the surrounding space and applying the right-hand rule shows the rotational direction of the space. Similarly, for a **negative point charge**, drawing rays from any point in space toward the negative charge shows the **opposite rotational direction**.

7. Geometric Explanation of Like Charges Repelling and Unlike Charges Attracting

Since charge is formed by the **cylindrical helical motion** of space around a particle, we can use this model to explain the behavior of charges, including why **like charges repel** and **opposite charges attract**.

When two **like charges** (both positive or both negative) come near each other, their cylindrical helical motion creates **increased spatial displacement** between them. The radial components of their motion move outward from each charge, causing an increase in the space between the charges, leading to **repulsion**. Additionally, the **rotational parts of the space** around the charges move in the same direction, causing a further increase in spatial quantity and reinforcing the **repulsive force**.

For **opposite charges**, the situation is different. The **cylindrical helical motion** of space around a positive charge radiates outward, while the space around a negative charge converges inward. When equal amounts of positive and negative charges meet, their **rotational motions cancel each other out**, and their **radial components** move at the speed of light from the positive charge toward the negative charge. This cancellation of the rotational components reduces the spatial quantity between the charges, causing **attraction**.

This explanation mirrors **Gauss's law for magnetism**, where the number of spatial displacement lines entering a surface equals the number exiting, resulting in a net cancellation. Similarly, the interaction between opposite charges leads to a complete cancellation of charge, resulting in neutralization.

When the charges are extremely close, their surrounding linear motions cancel each other out due to opposite directions, and their rotational motions cancel as well. This explains why, when equal amounts of positive and negative charges meet, the effects of the surrounding space motion disappear, leading to the **cancellation of the charges**.

Chapter 31. The Product of Velocity and the Rate of Change of Mass Over Time Represents the Electromagnetic Force

1. Electromagnetic Force in Unified Field Theory

In **Unified Field Theory**, the dynamics equation for force is expressed as:

$$\vec{F} = \frac{d\vec{p}}{dt} = \frac{d[m\,(\vec{c} - \vec{v})]}{dt} = \vec{c}\,\frac{dm}{dt} - \vec{v}\,\frac{dm}{dt} + m\,\frac{d\vec{c}}{dt} - m\,\frac{d\vec{v}}{dt}$$

where m is the mass of the particle, $\vec{c}$ is the **vector speed of light**, $\vec{v}$ is the velocity of the particle, and t is time.

The term $(\vec{c} - \vec{v})\,\frac{dm}{dt}$ represents the **force due to the change in mass**, called the **mass-acceleration force**. In **UFT**, this force is interpreted as the **electromagnetic force**, with the following breakdown:

- $\vec{c}\,\frac{dm}{dt}$ represents the **electric force**.

- $\vec{v}\,\frac{dm}{dt}$ represents the **magnetic force**.

2. Electromagnetic Force for a Stationary and Moving Charge

- **Static Electric Force**: In the s′ frame (rest frame), the system is at rest, with mass m′ and vector light speed $\vec{c}$. The **static electric force** is:

$$\overrightarrow{F_{rest}} = \vec{c'}\,\frac{dm'}{dt'}$$

- **Dynamic Electric Force**: In the s frame, where the system moves with velocity $\vec{v}$ along the x-axis, the surrounding space moves with light speed $\vec{c}$. The **dynamic electric force** is:

$$\overrightarrow{F_{X\,motion}} = \vec{c_x}\,\frac{dm}{dt}$$

And

$$\overrightarrow{F_{X\,rest}} = \vec{c'}_x\,\frac{dm'}{dt'}$$

Since dm/dt=dm'/dt', it follows that:

$$\overrightarrow{F_{X\,rest}} = \overrightarrow{F_{X\,motion}}$$

The scalar form of the force for the y and z components in the s frame can be expressed as:

$$F_y = c\sqrt{1 - \frac{v^2}{c^2}}\,\frac{dm}{dt}$$

This result aligns with the **relativistic transformation** of force components.

3. Electric and Magnetic Fields

In the **Unified Field Theory**, the **electric field** in the rest frame is:

$$\vec{E'} = \frac{\overrightarrow{F_{rest}}}{q} = \vec{c'}\,\frac{dm'}{dt'}\,\frac{1}{q}$$

The **electric field** in the moving frame is:

$$\vec{E} = \frac{\overrightarrow{F_{motion}}}{q} = \vec{c}\,\frac{dm}{dt}\,\frac{1}{q}$$

For components in the y and z axes, we have:

$$E'_y = E_y \sqrt{1 - \frac{v^2}{c^2}}$$

The relationship between the **electric field** $\vec{E}$ and the **magnetic field** $\vec{B}$ is given by:

$$\vec{B} = \frac{1}{c^2}\,\vec{v} \times \vec{E}$$

This result is consistent with **relativity**, where the magnetic field is a consequence of the motion of the charge relative to the observer.

4. Conclusion

The **Unified Field Theory** provides a framework in which the **electromagnetic force** is derived from the change in mass over time, breaking it into electric and magnetic components. The theory is consistent with the **relativistic transformations** of force and electric fields, showing that the electric and magnetic fields are interconnected through motion and relativistic effects.

Chapter 32. Definition Equation of the Nuclear Force Field

In **Unified Field Theory (UFT)**, all fields, including the **nuclear force field**, can be derived as variations of the **gravitational field**. Similar to the **electromagnetic field**, the nuclear force field can be expressed as a change in the gravitational field. However, the key difference is that the nuclear force field arises from the change in the **position vector** $\vec{r}$ (with magnitude r) of a point in space within the gravitational field over time, rather than from the mass change over time, as in the case of the electric field.

The **gravitational field** is expressed as:

$$\vec{A} = -G\,m\,\frac{\vec{r}}{r^3} = -G\,\frac{k}{\Omega}\,\frac{\vec{r}}{r^3}$$

where G is the gravitational constant, m is mass, $\vec{r}$ is the position vector, and Ω is the solid angle. The change in r^3 with respect to time t gives rise to the **nuclear force field**:

$$\vec{D} = -G\,m\,\frac{d\left(\frac{\vec{r}}{r^3}\right)}{dt} = -G\,m\,\frac{\left(\frac{d\vec{r}}{dt} - 3\,\frac{\vec{r}}{r}\,\frac{dr}{dt}\right)}{r^3} = -G\,m\,\frac{\left(\vec{c} - 3\,\frac{\vec{r}}{r}\,\frac{dr}{dt}\right)}{r^3}$$

Here, $\vec{c}$ represents the **vector of the speed of light**.

The above formula represents a hypothesis, as the **nuclear force field** is inherently different from the electric and magnetic fields. While humanity has developed formulas to describe electric and magnetic fields, the exact geometric nature of **charge** in these formulas remains unclear. Once the **geometric form of charge** is fully understood, we can integrate this into the electric and magnetic field equations, allowing UFT to express these fields in geometric terms.

However, the **nuclear force field** is more challenging to describe since no established formulas exist to account for nuclear forces in the same way that we describe electric and magnetic fields. **Nuclear forces** arise from interactions between **protons** and **neutrons** within the atomic nucleus, and since protons and neutrons are always in motion, even if the above nuclear force field formula is correct, it cannot be directly applied without accounting for **moving particles**.

The reliability of this nuclear force field formula, along with a precise formula for **nuclear interaction forces**, will require further theoretical and experimental investigation.

One hypothesis for the nuclear interaction force suggests that the **nuclear force** exerted by a particle of mass m on a nearby particle p (of mass m′) equals the nuclear force field $\vec{D}$ generated by the particle at point p, multiplied by the mass m′ of point p, or possibly crossed with the **momentum** m′v of point p, or the **angular momentum** $\vec{r} \times m' \vec{v}$.

Chapter 33. Magnetic Field Definition Equation

In **Unified Field Theory (UFT)**, the **magnetic field** and the **electric field** are fundamentally different types of fields that cannot directly interact or be superimposed. However, it is known that when a **charged particle** moves at a constant velocity relative to an observer, changes in the electric field occur. The portion of the electric field that changes can be interpreted as the **magnetic field**. In UFT, the **electric field** that varies with velocity is responsible for generating the **magnetic field**.

Consider a stationary particle at point o in an **inertial reference frame** s′, which has mass m′ and carries a **positive charge** q. This particle generates an **electrostatic field** $\vec{E'}$ at a point p in the surrounding space. The position vector from point o to point p is $\vec{r'}$, with a length r′. We can use this to define a **Gaussian surface** $s'=4\pi r'^2$ centered on o.

When the particle at point o moves with a velocity $\vec{v}$ along the x-axis in the reference frame s, the electric field changes due to this motion. The changing component of the electric field perpendicular to the velocity $\vec{v}$ is what we recognize as the **magnetic field** $\vec{B}$.

1. Magnetic Field and Cross-Product Relationship

The **magnetic field** $\vec{B}$ can be expressed as the product of the **moving electric field** $\vec{E}$ and the velocity $\vec{v}$:

$$\vec{B} = constant \times (\vec{v} \times \vec{E})$$

This relationship implies that the **magnetic field** is maximized when the electric field $\vec{E}$ is perpendicular to the velocity $\vec{v}$. Thus, the relationship between the electric and magnetic fields should be expressed as a **vector cross product**.

Using **Coulomb's law**, the **static electric field** $\vec{E}$ is:

$$\vec{E'} = \frac{q\,\vec{r'}}{4\pi\,\varepsilon_0\,r'^3}$$

Applying the **Lorentz transformation** (since the charge is moving relative to the observer), we obtain the **moving electric field** $\vec{E}$:

$$\vec{E} = \frac{q\,\gamma}{4\pi\,\varepsilon_0} \frac{(x - v\,t)\,\vec{\imath} + y\,\vec{\jmath} + z\,\vec{k}}{[\gamma^2\,(x - v\,t)^2 + y^2 + z^2]^{\frac{3}{2}}}$$

where $\gamma = \dfrac{1}{\sqrt{1-\frac{v^2}{c^2}}}$.

Thus, the **magnetic field** $\vec{B}$ becomes:

$$\vec{B} = \frac{\mu_0 q\,\gamma}{4\pi} \frac{\vec{v} \times \left[(x - v\,t)\,\vec{\imath} + y\,\vec{\jmath} + z\,\vec{k}\right]}{[\gamma^2\,(x - v\,t)^2 + y^2 + z^2]^{\frac{3}{2}}}$$

where μ_0 is the **vacuum permeability**. Using $\mu_0\varepsilon_0 = 1/c^2$, the equation simplifies to:

$$\vec{B} = \frac{1}{c^2}\,\vec{v} \times \vec{E}$$

2. Geometric Definition of the Magnetic Field

By incorporating the **geometric form** of **charge**

$$q = -k'\,k\,\frac{1}{\Omega^2}\frac{d\Omega}{dt}$$

, we can derive the geometric form of the **magnetic field**:

$$\vec{B} = -\frac{\mu_0 \, k' \, k}{4\pi} \frac{1}{\Omega^2} \frac{d\Omega}{dt} \, \gamma \, \frac{\vec{v} \times \left[(x - v\,t)\,\vec{\imath} + y\,\vec{\jmath} + z\,\vec{k}\right]}{[\gamma^2 \,(x - v\,t)^2 + y^2 + z^2]^{\frac{3}{2}}}$$

Let θ be the angle between the position vector $\vec{r}$ and the velocity $\vec{v}$, and let $r = \sqrt{\gamma^2 \,(x - v\,t)^2 + y^2 + z^2}$) . The **magnetic field** in **polar coordinates** can be expressed as:

$$\vec{B} = -\frac{\mu_0 \, k' \, k}{4\pi} \frac{1}{\Omega^2} \frac{d\Omega}{dt} \, \frac{v \, sin\theta}{\gamma^2 \, r^2 \,(1 - \beta^2 \, sin^2\theta)^{\frac{3}{2}}} \, \vec{e_r}$$

where $\beta = \frac{v}{c}$, v is the magnitude of the velocity, and $\vec{e_r}$ is the unit vector in the direction of $\vec{r}$.

3. Magnetic Field and Mass Relation

By using the relationship between mass and charge $= k' \frac{dm}{dt}$, we derive the equation incorporating **mass** into the magnetic field:

$$\vec{B} = \frac{\mu_0 \, k'}{4\pi} \frac{dm}{dt} \, \gamma \, \frac{\vec{v} \times \left[(x - v\,t)\,\vec{\imath} + y\,\vec{\jmath} + z\,\vec{k}\right]}{[\gamma^2 \,(x - v\,t)^2 + y^2 + z^2]^{\frac{3}{2}}}$$

4. Nature of Magnetic Fields and Right-Hand Rule

When a charged particle moves, the surrounding space rotates around the central axis defined by the velocity $\vec{v}$, and the **magnetic field** $\vec{B}$ follows the **right-hand rule**. This means that if you point your right hand's thumb in the direction of $\vec{v}$, your fingers curl in the direction of the **magnetic field** $\vec{B}$.

The relationship between the **magnetic field** B⃗, the moving **electric field** E⃗, and the velocity v⃗ is:

$$\vec{B} = \frac{1}{c^2}\ \vec{v} \times \vec{E}$$

This relationship follows the convention of vector cross products and **Stokes' theorem**, showing how the magnetic field components interact with the velocity and electric field

Chapter 34. Derivation of Maxwell's Equations

Maxwell's equations are fundamental for describing electromagnetic phenomena, but they are not the most fundamental principles in physics. By using the **definition equations** of electric and magnetic fields, along with **Gauss's theorem**, **Stokes's theorem** from field theory, and the **Lorentz transformation** from relativity, we can derive the four Maxwell equations.

1. Curl of the Static Electric Field $\overrightarrow{E'}$

For a stationary charge at point o, carrying a charge q, the **static electric field $\overrightarrow{E'}$** generated in the surrounding area can be described using the electric field definition equation:

$$\overrightarrow{E'} = f \, \frac{1}{\Omega^2} \, \frac{d\Omega}{dt} \, \frac{\vec{r}}{r^3}$$

Taking the **curl** of the field directly:

$$\nabla \times \overrightarrow{E'} = \vec{0}$$

In the above equation, only r^3 on the right-hand side is variable. The equation can be decomposed into three equalities:

$$\frac{\partial E_z'}{\partial y'} - \frac{\partial E_y'}{\partial z'} = 0$$

$$\frac{\partial E_x'}{\partial z'} - \frac{\partial E_z'}{\partial x'} = 0$$

$$\frac{\partial E_y'}{\partial x'} - \frac{\partial E_x'}{\partial y'} = 0$$

2. Divergence of the Static Electric Field $\overrightarrow{E'}$

Using the electric field definition equation:

$$\overrightarrow{E'} = f \, \frac{1}{\Omega^2} \frac{d\Omega}{dt} \frac{\vec{r}}{r^3}$$

Calculating the **divergence** directly:

$$\nabla \cdot \overrightarrow{E'} = 0$$

Here, r represents the radius of the Gaussian surface s surrounding point o. When r→0 (meaning point p on the Gaussian surface approaches charge o), the equation leads to an indeterminate form 0/0. Applying the **Dirac delta function**, we get:

$$\nabla \cdot \overrightarrow{E'} = \frac{\partial E'_x}{\partial x'} + \frac{\partial E'_y}{\partial y'} + \frac{\partial E'_z}{\partial z'} = \frac{\rho'}{\varepsilon_0}$$

where ρ' represents the **charge density** within the Gaussian surface s, and ε_0 is the vacuum permittivity constant. If point o is outside the Gaussian surface, the divergence remains zero.

3. Gaussian Law for a Moving Electric Field $\overrightarrow{E}$

Assume that the charge at point o is stationary in the s' frame, with charge q. In the s frame, this charge q moves at a constant velocity $\vec{v}$ along the x-axis. According to **relativity**, this motion causes **space contraction**, and the charge density ρ' increases by a factor of γ (the **Lorentz factor**):

$$\rho = \gamma \, \rho'$$

where $\gamma = \dfrac{1}{\sqrt{1-\frac{v^2}{c^2}}}$. Since the charge is moving, there is a current density:

$$\vec{J} = \rho\, v\, \vec{\imath} = \gamma\, \rho'\, v\, \vec{\imath}$$

From the **Lorentz transformation**, we get:

$$x' = \gamma\, (x - v\, t)$$

and from the **relativistic transformation** of the electric field:

$$E_x = E'_x \,, \quad E_y = \gamma\, E'_y \,, \quad E_z = \gamma\, E'_z$$

The **Gaussian law** for the moving electric field $\vec{E}$ becomes:

$$\nabla \cdot \vec{E} = \frac{\partial E_x}{\partial x} + \frac{\partial E_y}{\partial y} + \frac{\partial E_z}{\partial z} = \frac{\rho}{\varepsilon_0}$$

4. Gaussian Law for the Magnetic Field

Using the differential operators $\dfrac{\partial}{\partial y} = \dfrac{\partial}{\partial y'}$, $\dfrac{\partial}{\partial z} = \dfrac{\partial}{\partial z'}$, and based on the relationship between the **magnetic field** $\vec{B}$ and the **electric field** $\vec{E}$, we have:

$$B_x = 0$$

$$B_y = \frac{v}{c^2}\, E_z$$

$$B_z = -\frac{v}{c^2}\, E_y$$

Applying the **curl** of the electrostatic field $\vec{E}$:

$$\nabla \cdot \overrightarrow{E'} = 0$$

and using the **relativistic transformation** of the electric field:

$$\gamma \, E_z' = E_z \,, \qquad \gamma \, E_y' = E_y$$

we derive **Gauss's law for magnetism**:

$$\nabla \cdot \vec{B} = 0$$

5. Faraday's Law of Electromagnetic Induction

Starting from the **curl** of the electrostatic field $\overrightarrow{E'}$, and using the relativistic transformation for Ez and Ey:

$$\frac{1}{\gamma} \left(\frac{\partial E_z}{\partial y} - \frac{\partial E_y}{\partial z} \right) = 0$$

we obtain:

$$\frac{\partial E_z}{\partial y} - \frac{\partial E_y}{\partial z} = 0$$

From the second curl equation:

$$\frac{\partial E_x'}{\partial z'} - \frac{\partial E_z'}{\partial x'} = 0$$

Using the **Lorentz transformation** $x' = \gamma \, (x - v \, t)$, and applying the chain rule for partial derivatives, we derive:

$$\frac{\partial E_x}{\partial z} - \frac{\partial E_z}{\partial x} = -\frac{v^2}{c^2} \frac{\partial E_z}{\partial t}$$

From the relationship between the **magnetic field** and the **electric field**:

$$B_y = \frac{v}{c^2} \, E_z$$

we get **Faraday's law**:

$$\nabla \times \vec{E} = -\frac{\partial \vec{B}}{\partial t}$$

6. Ampère's Law (with Maxwell's Correction)

Finally, using the same relationships between the **magnetic field** $\vec{B}$ and the **electric field** $\vec{E}$, and applying **Stokes's theorem**, we derive:

$$\nabla \times \vec{B} = \mu_0 \vec{J} + \mu_0 \, \varepsilon_0 \, \frac{\partial \vec{E}}{\partial t}$$

where $\vec{J}$ is the current density, and ε_0 and μ_0 are the vacuum permittivity and permeability constants, respectively.

Chapter 35. Gravitational Field Varying Over Time Produces an Electric Field

In Unified Field Theory, the gravitational field is considered the primary field, from which the electric field, magnetic field, and nuclear force field are derived through variations in the gravitational field. Charge originates from variations in mass. Conversely, changes in the electric, magnetic, and nuclear force fields can also generate a gravitational field. However, the forms of these changes are more complex, whereas the process by which a gravitational field produces other fields is relatively simpler.

1. Electric Field from a Stationary Particle's Gravitational Field

We begin by calculating the electric field produced by a varying gravitational field when the particle at point o is stationary relative to the observer. Using the gravitational field equation:

$$\vec{A} = -\frac{G\,m'\,\vec{r}}{r^3} = -G\,k\,\frac{1}{\Omega}\frac{\vec{r}}{r^3}$$

where $\vec{r}$ is the position vector and Ω is the solid angle. Taking the partial derivative of $1/\Omega$ with respect to time t, we obtain:

$$\frac{\partial \vec{A}}{\partial t} = G\,k\,\frac{1}{\Omega^2}\frac{d\Omega}{dt}\frac{\vec{r}}{r^3}$$

From the geometric definition of the electrostatic field:

$$\vec{E} = -\frac{k'}{4\pi\,\varepsilon_0}\frac{1}{\Omega^2}\frac{d\Omega}{dt}\frac{\vec{r}}{r^3}$$

we can relate this to the gravitational field by:

$$\vec{E} = -\frac{k'}{4\pi \, \varepsilon_0 \, G} \frac{d\vec{A}}{dt}$$

Since G, k', 4π, and ε_0 are constants, we combine them into a constant f, giving:

$$\vec{E} = -f \frac{d\vec{A}}{dt}$$

This leads to the following component equations:

$$E_x = -f \frac{\partial A_x}{\partial t}$$

$$E_y = -f \frac{\partial A_y}{\partial t}$$

$$E_z = -f \frac{\partial A_z}{\partial t}$$

2. Electric Field from a Moving Particle's Gravitational Field

When a charged particle at point o moves with constant velocity $\vec{v}$ (with scalar value v) along the positive x-axis relative to the observer, we use relativistic transformations to derive the relationship between the electric field and gravitational field.

For a stationary particle, the electric and gravitational fields are related by:

$$E'_x = -f \frac{\partial A'_x}{\partial t'}$$

$$E_y' = -f \, \frac{\partial A_y'}{\partial t'}$$

$$E_z' = -f \, \frac{\partial A_z'}{\partial t'}$$

From the Lorentz transformation for electric fields in relativity:

$$E_x = E_x' , \quad E_y = \gamma \, E_y' , \quad E_z = \gamma \, E_z'$$

where $\gamma = \dfrac{1}{\sqrt{1-\frac{v^2}{c^2}}}$.

For the relativistic transformation of the gravitational field:

$$A_x = \gamma \, A_x' , \quad A_y = \gamma^2 \, A_y' , \quad A_z = \gamma^2 \, A_z'$$

Using the Lorentz time transformation $t' = \gamma \left(t - \frac{v \, x}{c^2} \right)$, we differentiate with respect to time:

$$\frac{\partial t'}{\partial t} = \gamma \left(\frac{\partial t}{\partial t} - \frac{v^2}{c^2} \right) = \frac{\gamma}{\gamma^2} = \frac{1}{\gamma}$$

Thus:

$$\frac{\partial}{\partial t'} = \gamma \, \frac{\partial}{\partial t}$$

3. Relationship Between Electric and Gravitational Fields for a Moving Object

From this, we derive the relationship between the moving electric field $\vec{E}$ and the moving gravitational field $\vec{A}$:

$$E_x = -f \, \frac{\partial A_x}{\partial t}$$

$$E_y = -f \, \frac{\partial A_y}{\partial t}$$

$$E_z = -f \, \frac{\partial A_z}{\partial t}$$

These results show that the relationship between the electric field and the gravitational field remains consistent whether the particle is stationary or moving at a constant velocity.

Chapter 36. The Change in the Gravitational Field of a Uniformly Moving Object Generates an Electric Field

As discussed earlier, when the particle at point o is stationary relative to an observer, the divergence of the surrounding gravitational field $\overrightarrow{A'}$ is given by:

$$\nabla \cdot \overrightarrow{A'} = \frac{\partial A'_x}{\partial x'} + \frac{\partial A'_y}{\partial y'} + \frac{\partial A'_z}{\partial z'}$$

where A'_x , A'_y , A'_z are the components of $\overrightarrow{A'}$ along the three coordinate axes.

When point o moves with constant velocity $\vec{v}$ (with magnitude v) along the positive x-axis relative to the observer, the divergence of the gravitational field $\vec{A}$ becomes:

$$\nabla \cdot \vec{A} = \frac{\partial A_x}{\partial x} + \frac{\partial A_y}{\partial y} + \frac{\partial A_z}{\partial z}$$

Using the Lorentz transformations $x' = \gamma\,(x - v\,t)$ and differentiating, we obtain:

$$\frac{1}{\gamma}\frac{\partial}{\partial x} = \frac{\partial}{\partial x'}$$

then add $\partial y = \partial y'$, $\partial z = \partial z'$,

Applying these transformations and the relativistic transformation of the gravitational field, we have:

$$\nabla \cdot \overrightarrow{A'} = \frac{1}{\gamma^2} \frac{\partial A_x}{\partial x} + \frac{1}{\gamma^2} \frac{\partial A_y}{\partial y} + \frac{1}{\gamma^2} \frac{\partial A_z}{\partial z}$$

This can be rewritten as:

$$\nabla \cdot \vec{A} = \left(1 - \frac{v^2}{c^2} \right) \nabla \cdot \vec{A}$$

where v is the velocity of the particle, and c is the speed of light.

Rewriting the equation in vector form, since we are dealing with a divergence, not a curl, we apply the dot product of the velocity vector $\vec{v}$ (along the x direction) and the components of the gravitational field $\vec{A}$:

$$\nabla \cdot \overrightarrow{A'} = \left(1 - \frac{v^2}{c^2} \right) \nabla \cdot \vec{A}$$

Now, using the relation between the gravitational field component Ax and the electric field component Ex, where:

$$E_x = -f \frac{\partial A_x}{\partial t}, \quad \text{and } v \frac{\partial}{\partial x} = \frac{\partial}{\partial t}$$

we can rewrite the equation as:

$$\nabla \cdot \overrightarrow{A'} = \nabla \cdot \vec{A} - \frac{v}{c^2} \left(\frac{\partial A_x}{\partial t} \right) = \nabla \cdot \vec{A} + \frac{v}{c^2 f} E_x$$

This indicates that when the particle at point o is stationary, it generates a gravitational field $\overrightarrow{A'}$ in the surrounding space. When the particle moves with velocity $\vec{v}$, the gravitational field changes, splitting into two parts: one independent of velocity and the other dependent on velocity. The velocity-dependent part along the x-axis is effectively the electric field.

By relating the gravitational field and electric field of a moving particle, we can also derive the relationship between the curl of the magnetic field and the changing gravitational field.

Using the relationship between the moving electric field $\vec{E}$ and the moving gravitational field $\vec{A'}$:

$$\vec{E} = -f\,\frac{d\vec{A}}{dt}$$

and substituting it into Maxwell's equations:

$$\mu_0 \vec{J} + \frac{1}{c^2}\frac{\partial \vec{E}}{\partial t} = \nabla \times \vec{B}$$

we obtain:

$$\mu_0 \vec{J} - \frac{f}{c^2}\frac{\partial^2 \vec{A}}{\partial t^2} = \nabla \times \vec{B}$$

where $\vec{J}$ is the current formed by the moving charge with density ρ, and

$$\mu_0 \vec{J} = \frac{\vec{v}\,\rho}{c^2}$$

Therefore:

$$f\frac{\partial^2 \vec{A}}{\partial t^2} = \frac{v}{fc^2}\left(\nabla \cdot \vec{E}\right) - c^2\,\nabla \times \vec{B}$$

This equation demonstrates that a changing gravitational field can generate both an electric field and a magnetic field, extending the Maxwell equations to include gravitational effects.

Chapter 37. Gravitational Field Generated by a Moving Charge

1. Gravitational Field Generated by the Magnetic Field of a Charge in Uniform Linear Motion

Figure 4: Gravitational Field Generated by Moving Charge

The core principle of Unified Field Theory is that a changing gravitational field can generate an electric field, and conversely, a changing electromagnetic field can generate a gravitational field. According to relativity and electromagnetism, a moving charge generates not only an electric field but also a magnetic field.

Unified Field Theory extends this by positing that a moving charge also generates a gravitational field. Below, we derive the relationship between the electromagnetic field and the gravitational field produced by a moving charge.

Previously, we discussed that the electric field generated by a changing gravitational field does not change direction; the gravitational and electric fields align. In general, the electric field is always perpendicular to the magnetic field, and therefore, the gravitational field is typically perpendicular to the magnetic field as well.

We will explore the relationship between the curl of the gravitational field and the magnetic field, as the curl describes field variation in the perpendicular direction, while divergence describes field variation in the parallel direction.

Consider a point charge at point o, which at time t=0 starts from the origin and moves in uniform linear motion at speed v along the positive x-axis relative to the observer. The point charge at o generates an electric field $\vec{E}$, a magnetic field $\vec{B}$, and a gravitational field $\vec{A}$ at point p in the surrounding space.

To analyze this, we focus on the spatial point p as our observation point. The rotational directions of the gravitational field $\vec{A}$ and the electric field $\vec{E}$, are consistent, both forming a left-hand spiral. However, at any specific point on this rotational line, $\vec{A}$ and $\vec{E}$, are perpendicular to each other.

To prove that the electric field $\vec{E}$, the magnetic field $\vec{B}$, and the gravitational field $\vec{A}$ satisfy the relationship shown, we first calculate the curl (rotational) of $\vec{A}$:

$$\nabla \times \vec{A} = \left(\frac{\partial A_z}{\partial y} - \frac{\partial A_y}{\partial z} \right) \vec{\imath} + \left(\frac{\partial A_x}{\partial z} - \frac{\partial A_z}{\partial x} \right) \vec{\jmath} + \left(\frac{\partial A_y}{\partial x} - \frac{\partial A_x}{\partial y} \right) \vec{k}$$

From previous analysis, the curl of the gravitational field when the object is stationary is zero:

$$\nabla \times \vec{A'} = 0$$

In component form, this is:

$$\frac{\partial A'_z}{\partial y'} - \frac{\partial A'_y}{\partial z'} = 0$$

$$\frac{\partial A'_x}{\partial z'} - \frac{\partial A'_z}{\partial x'} = 0$$

$$\frac{\partial A'_y}{\partial x'} - \frac{\partial A'_x}{\partial y'} = 0$$

By applying the relativistic transformation of the gravitational field, we obtain:

$$0 = \frac{\partial A'_z}{\partial y'} - \frac{\partial A'_y}{\partial z'} = \frac{1}{\gamma^2} \left(\frac{\partial A_z}{\partial y} - \frac{\partial A_y}{\partial z} \right)$$

where $\gamma = \frac{1}{\sqrt{1-\frac{v^2}{c^2}}}$ is the relativistic factor. $\partial y' = \partial y$, $\partial z' = \partial z$

Thus, the curl simplifies to:

$$\frac{\partial A_z}{\partial y} - \frac{\partial A_y}{\partial z} = 0$$

Taking the partial derivative of the Lorentz transformation, $x' = \gamma (x - v\,t)$, we get:

$$\frac{1}{\gamma}\frac{\partial}{\partial x} = \frac{\partial}{\partial x'}$$

From:

$$\frac{\partial A'_x}{\partial z'} - \frac{\partial A'_z}{\partial x'} = 0$$

we obtain:

$$\frac{1}{\gamma}\frac{\partial A_x}{\partial z} - \frac{1}{\gamma^3}\frac{\partial A_z}{\partial x} = 0$$

This simplifies to:

$$\frac{\partial A_x}{\partial z} - \left(1 - \frac{v^2}{c^2}\right)\frac{\partial A_z}{\partial x} = 0$$

We continue analyzing the remaining curl components in a similar fashion. For the y- and x-components, we similarly apply relativistic transformations and obtain the following relationships:

$$\frac{\partial A_y}{\partial x} - \frac{\partial A_x}{\partial y} = \frac{v^2}{c^2}\frac{\partial A_y}{\partial x}$$

Combining the curl components gives us the general form:

$$\vec{\nabla} \times \vec{A} = \frac{1}{f}\vec{B}$$

where f is a proportionality factor that depends on the properties of the fields and the relativistic effects in play.

Thus, we arrive at the fundamental relationship between the curl of the gravitational field $\vec{A}$ and the magnetic field $\vec{B}$. This equation reveals that the magnetic field can be expressed as the curl of the gravitational field.

At a given moment, the magnetic, electric, and gravitational fields are mutually perpendicular.

This relationship may provide insight into the AB effect in quantum mechanics.

By integrating both sides using Stokes' theorem, we derive the following integral equation:

$$\oint \vec{A} \cdot d\vec{l} = \frac{1}{f} \oiint \vec{B} \cdot d\vec{S}$$

2. A Time-Varying Magnetic Field Generates an Electric Field and a Gravitational Field

Imagine a point charge at point o that, at time t=0, starts moving in a uniform straight line along the positive x-axis at a constant speed $\vec{v}$ (magnitude v) relative to us as observers. This point charge generates a moving electric field $\vec{E}$ and a uniform magnetic field $\vec{B}$ at any arbitrary space point p around it:

$$\vec{B} = \frac{1}{c^2} \vec{v} \times \vec{E}$$

Now, suppose the charge at point o accelerates in the positive x-axis direction, with an acceleration $\vec{A}$ (magnitude a) relative to us. This acceleration causes the charge to generate a moving electric field $\vec{E}$, a time-varying magnetic field $\frac{d\vec{B}}{dt}$, and a gravitational field $\vec{A}$ at any arbitrary space point p around it.

To analyze this scenario, we take the spatial point p as the observation point. By differentiating the equation $\vec{B} = \frac{1}{c^2} \vec{v} \times \vec{E}$ with respect to time, we obtain:

$$\frac{d\vec{B}}{dt} = \frac{1}{c^2} \frac{d\vec{v}}{dt} \times \vec{E} + \frac{1}{c^2} \vec{v} \times \frac{d\vec{E}}{dt}$$

If we prove that $\frac{d\vec{B}}{dt} = \frac{1}{c^2} \vec{v} \times \frac{d\vec{E}}{dt}$ corresponds to Faraday's electromagnetic induction principle (where a changing magnetic field produces a changing electric field), we can similarly conclude that:

$$\frac{d\vec{B}}{dt} = \frac{1}{c^2} \frac{d\vec{v}}{dt} \times \vec{E}$$

represents a changing magnetic field generating a gravitational field.

Since $\frac{d\vec{v}}{dt} = \vec{A}$ is the acceleration of the space point p, and according to Unified Field Theory, the acceleration of space itself is equivalent to a gravitational field, we can first demonstrate that:

$$\frac{d\vec{B}}{dt} = \frac{1}{c^2} \vec{v} \times \frac{d\vec{E}}{dt}$$

represents the Faraday electromagnetic induction principle.

Next, by applying the relationship between $\vec{B}$ and $\vec{E}$, we obtain:

$$\frac{d\vec{B}}{dt} = \frac{1}{c^2} \vec{v} \times \frac{d\vec{E}}{dt}$$

These equations match the components of the electromagnetic induction principle:

$$\nabla \times \vec{E} = -\frac{\partial \vec{B}}{\partial t}$$

We can now analyze the generation of the gravitational field $\vec{A}$ by the changing magnetic field $\vec{B}$:

$$\frac{\partial \overrightarrow{B_x}}{\partial t} = \vec{0}$$

$$\frac{\partial \overrightarrow{B_y}}{\partial t} = \frac{1}{c^2}\frac{\partial \vec{v}}{\partial t} \times \overrightarrow{E_z} = \frac{1}{c^2}\vec{A} \times \overrightarrow{E_z}$$

$$\frac{\partial \overrightarrow{B_z}}{\partial t} = -\frac{1}{c^2}\frac{\partial \vec{v}}{\partial t} \times \overrightarrow{E_y} = -\frac{1}{c^2}\vec{A} \times \overrightarrow{E_y}$$

Thus, we can express the relationship between the time-varying magnetic field and the gravitational field as:

$$\frac{d\vec{B}}{dt} = \frac{1}{c^2}\vec{A} \times \vec{E}$$

In summary, an accelerating positive charge generates a gravitational field in the surrounding space, opposite to the direction of acceleration, and this gravitational field propagates outward at the speed of light. The equations for the changing electromagnetic field generating a gravitational field are consistent with the fundamental relationship

$$\vec{B} = \frac{\vec{v}}{c^2}\vec{E}$$

3. The Relationship Between the Electric Field, Magnetic Field, and Gravitational Field of an Accelerating Charge

The creation of a gravitational field by a changing electromagnetic field is a central concept of Unified Field Theory and forms the basis for the application of artificial field technology. To further explore this, we will derive the gravitational field generated by an accelerating positive charge using an alternative approach.

The relationships between the electric field, magnetic field, and gravitational field can be seen as extensions of the basic equation for the magnetic field:

$$\vec{B} = \frac{\vec{v}}{c^2}\,\vec{E}$$

All of the relationships between these fields are derived from this fundamental equation. For example, the equation:

$$dB/dt = A \times E/c^2$$

applies specifically to microscopic elementary particles. In the macroscopic view, particles consist of many smaller charged particles, with positive and negative charges largely canceling each other out, as do many magnetic fields. Therefore, the gravitational field produced by a changing magnetic field may only apply to positive charges. This is because the divergent movement of space around a positive charge propagates the space distortion effects (including the gravitational field formed by accelerated electric and magnetic fields) outward at the speed of light.

However, the inward convergent movement of space around a negative charge theoretically cannot propagate the space distortion effects outward. According to Lorentz transformation, the light-speed movement causes space to contract to zero, effectively making it no longer part of the observable space, leading to uncertainty. Whether this formula applies to negative charges still requires further theoretical exploration and practical testing.

To better understand the relationship between the electric field, magnetic field, and gravitational field of an accelerating charge, let's analyze an example.

1) Example: Accelerating Positive Charge

Consider a stationary point charge o with a positive charge q relative to an observer, generating a static electric field $\vec{E}$ at a point p in the surrounding space. At time t=0, point o accelerates with vector acceleration $\vec{G}$ (magnitude g) along the positive x-axis.

According to Unified Field Theory, the acceleration of point o will cause the space point p to move outward from point o at light speed C, superimposing an acceleration $-\vec{G}$. Since the gravitational field is defined as the acceleration of space points, the gravitational field $\vec{A}$ (magnitude a) at point p is equivalent to the acceleration $-\vec{G}$. Therefore, point p experiences a gravitational field due to the acceleration of point o:

$$A = -G$$

Next, we determine the relationship between the static electric field , the transverse electric field $\vec{E\theta}$ due to acceleration, and the gravitational field $\vec{A}$.

Assume point o accelerates in a straight line with uniform acceleration $\vec{G}$. At time t=τ, point o reaches point d and stops accelerating, with velocity v=gτ. It then continues to move uniformly along the x-axis at velocity v until reaching point q.

Now, consider the distribution of the electric field around charge o at time t (where t≫τ). Between time 0 and time τ, the acceleration of charge o distorts the electric field lines, which extend outward at the speed of light c. Unified Field Theory states that electric field lines are displacements of space points around the charge, moving at the speed of light.

The distortion caused by the acceleration of charge o extends outward like a faucet spraying water uniformly. When the faucet shakes, the water flow distorts, and this distortion propagates outward at the speed of the water flow—similarly, the distortion of the electric field lines propagates outward at light speed. The distorted region has a thickness $c\tau$, and is sandwiched between two spherical surfaces.

At time t, the back surface has propagated outward a distance $c(t-\tau)$, centered at point q, while the front surface has propagated outward a distance ct, centered at point o. The electric field within this spherical surface is that of a uniformly moving charge. Given that $v \ll c$, the electric field within the spherical surface can be approximated as a static electric field.

2) *Field Components*

Within the distorted region, the electric field can be divided into two components: the radial electric field $\vec{E_r}$, and the transverse electric field $\vec{E_\theta}$, caused by the acceleration. The relationship between these components is:

$$E\theta/Er = g\ r\ \sin\theta/c^2$$

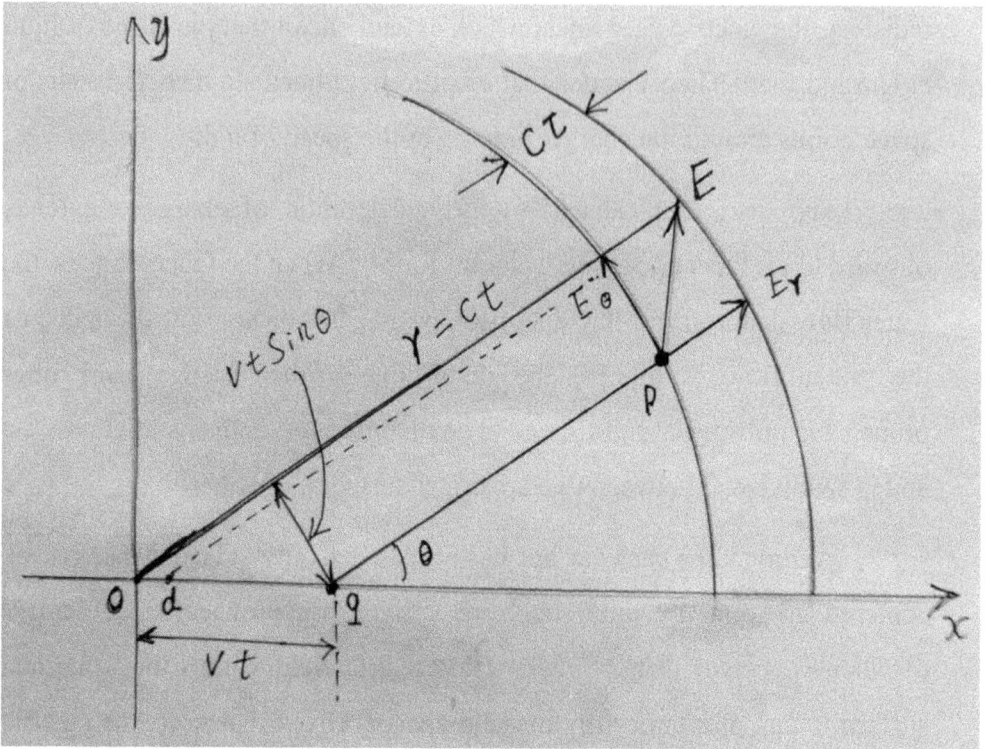

In Unified Field Theory, the gravitational field is the acceleration of space points and is directed opposite to the position vector $\vec{R}$ pointing from the source of the gravitational field to the observation point. Therefore, the gravitational field $\vec{A}$ can be expressed as:

$$E\theta/Er = \vec{A} \times \vec{R}/c2$$

Here, $\vec{R} = ct$ is the vector from point o to the observation point p.

3) *Magnetic Field*

Next, we derive the relationship between the changing magnetic field and the gravitational field produced by the accelerating charge. According to Maxwell's equations, a changing electric field generates a changing magnetic field. The relationship between the transverse electric field $E\theta^{\to}$ and the magnetic field $B\theta^{\to}$ is:

$$B\theta = C \times E\theta / c^2$$

where $C^{\to}$ is the vector light speed. Since the charge's velocity v is much smaller than c, the speed of the space point p can still be considered vector light speed $C^{\to}$. Therefore:

$$cb\theta = e\theta$$

Using this relationship and the spacetime unification equation R=Ct, we can express the magnetic field as:

$$B\theta / Er = A \times R / c^3$$

Thus, the relationship between the original electric field $Er^{\to}$, the gravitational field $A^{\to}$, and the changing magnetic field $B\theta^{\to}$ is:

$$B\theta = (A \times Er)t / c^2$$

Taking the time derivative, we get:

$$dB\theta / dt = A \times Er / c^2$$

This matches the magnetic field definition equation $B = V \times E / c^2$, differentiated with respect to time:

$$dB / dt = A \times E / c^2$$

4. Conclusion

An accelerating positive charge generates a gravitational field in the surrounding space that is opposite to the direction of acceleration and propagates outward at the speed of light. The changing electromagnetic field produces both a changing electric field and a gravitational field, consistent with the fundamental relationship $B = V \times E/c^2$. All relationships between the electric field, magnetic field, and gravitational field can be derived as variations of this fundamental equation.

Chapter 38. Experimental Situations of Gravitational Fields Generated by Changing Electromagnetic Fields

A patent titled *"A Device for Electromagnetic Conversion to Gravitational Fields"* has been registered to demonstrate gravitational fields generated by changing electromagnetic fields. These *artificial fields* can be controlled and may eventually replace electric power, ushering humanity into an era of light-speed virtual technology.

专利受理通知书+一种电磁转化引力场...　...
文件预览

国 家 知 识 产 权 局

100000

北京市房山区良乡凯旋大街建设路 18 号－1322512（集群注册）　北京思专专利代理事务所(普通合伙)

发文日：

2024 年 03 月 07 日

申请号：202410262397.5　　　　　发文序号：2024030701252710

专 利 申 请 受 理 通 知 书

根据专利法第 28 条及其实施细则第 43 条、第 44 条的规定，申请人提出的专利申请已由国家知识产权局受理。现将确定的申请号、申请日等信息通知如下：

申请号：202410262397.5
申请日：2024 年 03 月 07 日
申请人：张祥朋
发明人：张祥朋
发明创造名称：一种电磁转化引力场装置
经核实，国家知识产权局确认收到文件如下：
权利要求书 1 份 1 页，权利要求项数： 3 项
说明书 1 份 4 页
说明书附图 1 份 2 页
说明书摘要 1 份 1 页
专利代理委托书 1 份 2 页
发明专利请求书 1 份 4 页
实质审查请求书 文件份数：1 份

提示：
1 申请人收到专利申请受理通知书之后，认为其记载的内容与申请人所提交的相应内容不一致时，可以向国家知识产权局请求更正。
2 申请人收到专利申请受理通知书之后，向向国家知识产权局办理各种手续时，均应写清楚、清晰地写明申请号。

审 查 员：自动受理
联系电话：010-62356655

审查部门：初审及流程管理部

初审合格+一种电磁转化引力场装置+... ...

文件预览

国 家 知 识 产 权 局

100000

发文日：

2024年06月05日

申请号或专利号：202410262397.5　　　发文序号：

申请人或专利权人：

发明创造名称：一种电磁转化引力场装置

发 明 专 利 申 请 初 步 审 查 合 格 通 知 书

The key to realizing artificial fields lies in the successful experimentation with generating gravitational fields from changing electromagnetic fields. On November 2, 2023, I discovered that accelerating positive charges produce a gravitational field in the opposite direction of their acceleration. Later, on March 1, 2024, I found that changing magnetic fields generate vortex gravitational fields, which cause objects to rotate.

1. Experiment: Accelerating Positive Charges Producing a Linear Gravitational Field Opposite to the Acceleration Direction

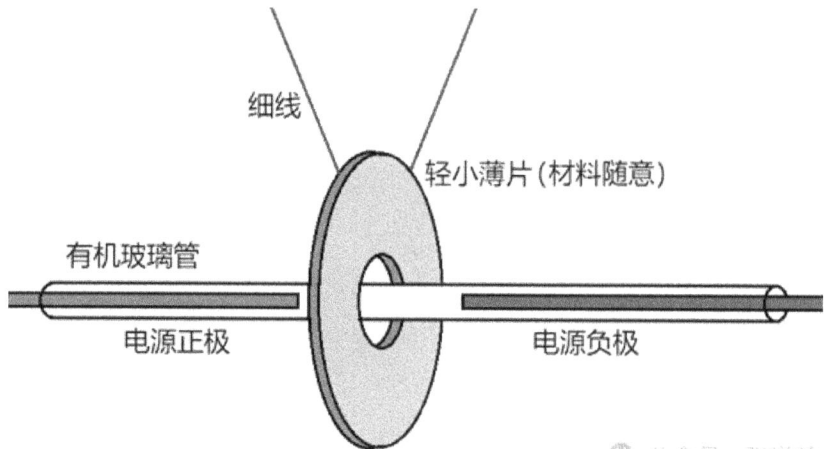

In the experimental setup, a high-voltage DC power source (over 30,000 volts) is connected to terminals placed 6 cm apart in an acrylic tube. A lightweight, thin, sheet-like object is suspended on a string, with a hole in its center, positioned in the middle of the gap between the positive and negative terminals. Upon pressing the power switch, the suspended object moves toward the positive terminal, regardless of polarity. This phenomenon is due to the generation of a gravitational field by accelerating positive charges, as described by the equation:

$$E\theta/Er = A \times R/ c^2.$$

where $E\theta$ is the distorted electric field generated by the accelerated charge, Er is the static electric field when the charge is stationary, R is the position vector, A is the gravitational field, and c is the speed of light. The gravitational field causes the suspended object to accelerate. To minimize interference from ion wind and electrostatic effects, the acrylic tube encloses the wire, and the suspended object is designed to be flat and thin to reduce polarization effects.

In a modified version of the experiment, a flexible silicone tube was used instead of the acrylic tube to prevent the generation of ion wind from the wire connections, achieving better results.

Equipment Setup:

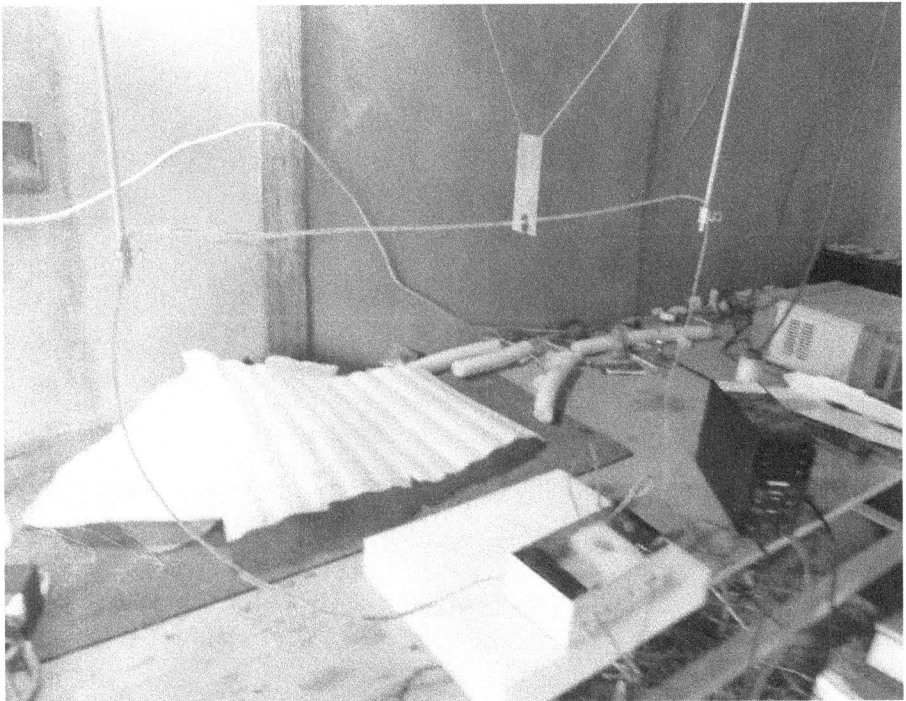

- A 190 cm long silicone tube with an outer diameter of 3 mm (or 2 mm) and an inner diameter of 1 mm.

- Two 90 cm long enameled copper wires (0.8 mm diameter) threaded through the silicone tube, spaced 4.5 cm apart.

- A lightweight plastic sheet (4 cm x 11 cm, 0.15 mm thick) suspended at the center of the two wires, with a hole in its center to fit around the silicone tube.

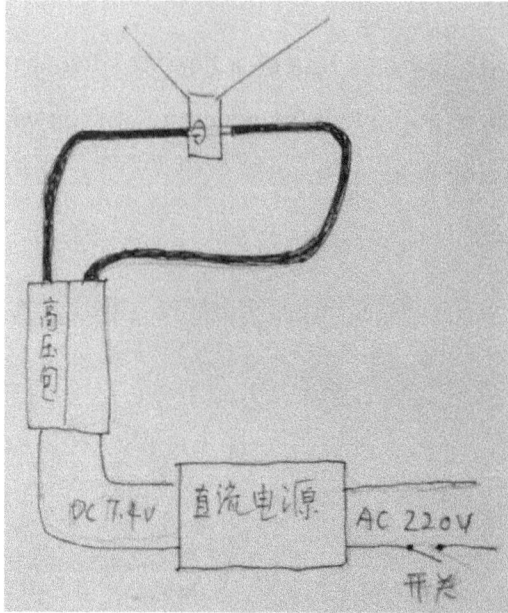

When powered by two series-connected high-voltage packs (each generating about 30 kV), the plastic sheet moves towards the positive terminal upon pressing the power switch. This experiment, when conducted under vacuum conditions, confirms that the gravitational field generated by the accelerated positive charges is responsible for the movement.

2. Experiment: Vortex Gravitational Field Generated by Changing Magnetic Fields

In this experiment, two spiral coils (19 cm long, 3.7 cm in diameter) made of 0.57 mm diameter enameled copper wire are wound around a 1 mm thick paper tube. The top coil is connected to the negative terminal, while the bottom coil is connected to the positive terminal of a high-voltage power source. Both coils are placed around a vacuum chamber containing a small red polyethylene ball suspended by a thread.

When the power is turned on, the polyethylene ball rotates inside the vacuum chamber, with the rotation axis aligned with the magnetic field lines. This demonstrates the generation of a vortex gravitational field by the changing magnetic field. The vacuum environment rules out ion wind and electrostatic motor effects, as there is no air to ionize, and the electrodes are outside the vacuum chamber.

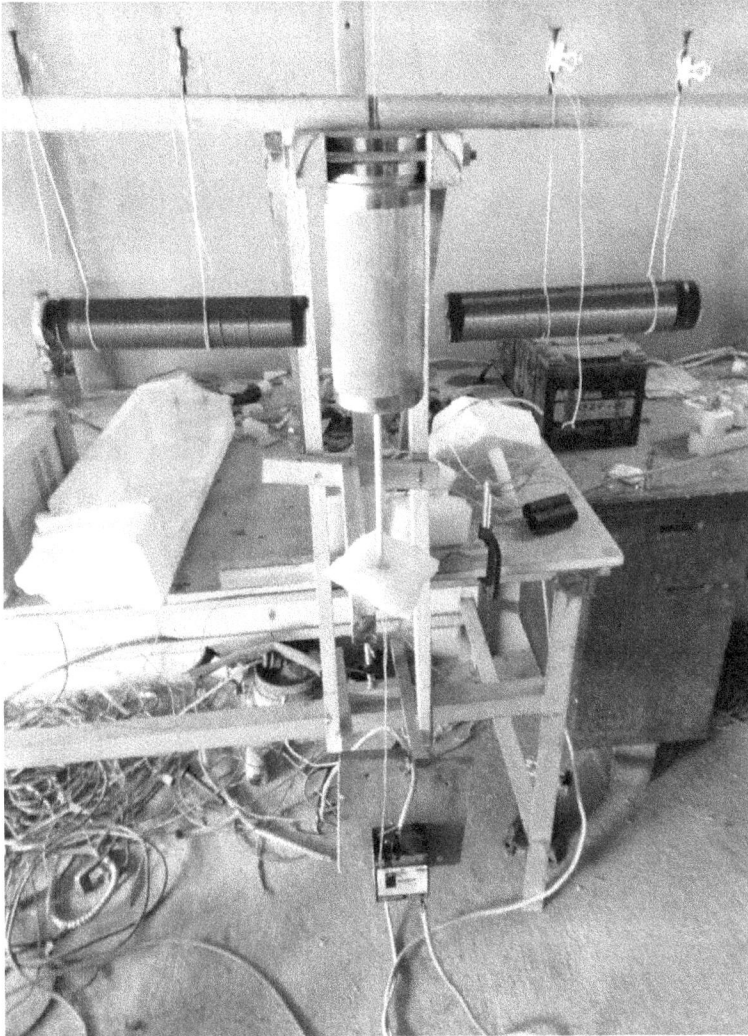

Although polarization effects still exist in a vacuum, their influence on rotation is minimized. The coils are aligned in such a way that the gravitational field generated by the changing magnetic field becomes the dominant effect, and polarization's impact on rotation speed can be dismissed.

Further testing using a Faraday cage confirms that when the power is turned on and off, the polyethylene ball inside continues to rotate, even after the power is off. This effect supports Unified Field Theory's prediction that a changing magnetic field generates a vortex gravitational field.

3. Conclusion

These experiments illustrate the principles of gravitational fields generated by changing electromagnetic fields, as predicted by Unified Field Theory. Both linear and vortex gravitational fields can be produced by accelerating charges and changing magnetic fields, respectively. These fields offer the potential for groundbreaking applications in artificial fields, moving toward the future of controlled gravitational technologies.

Chapter 39. Unified Field Theory Energy Equation

1. Energy Definition

Energy measures the motion of a particle in space (or the space surrounding the particle) relative to an observer within a certain spatial range. Due to the unification of space and time, this can also be understood as occurring within a certain period.

The definition of energy, like that of momentum, reflects the degree of motion of the particle and space relative to the observer. The key difference is that momentum is a vector, while energy is a scalar. Momentum describes the direction of motion, while energy quantifies the magnitude of motion.

It is crucial to note that energy depends on four essential elements: space, a material point (particle), an observer, and motion. If any of these elements are missing, the concept of energy loses meaning. For instance, empty space (pure vacuum) without objects has no energy. Additionally, without specifying an observer, energy cannot be determined.

2. Unified Field Theory Energy Equation

Multiplying both sides of the scalar form of the Unified Field Theory momentum equation $m'c = mc\sqrt{(1 - v^2/c^2)}$ by the scalar speed of light c, we obtain the Unified Field Theory energy equation:

$$e = m'c^2 = mc^2\sqrt{(1 - v^2/c^2)}$$

Here, $m'c^2$ is the energy of point o when at rest, which is consistent with the view in relativity. For a particle at rest relative to the observer, relativity gives a rest energy $E = m'c^2$, implying that the particle is surrounded by vector light-speed paths, determined by the mass m'.

Unified Field Theory assumes that any object at rest has space radiating outward at light speed, directly explaining relativistic rest energy. According to Unified Field Theory, the energy of a moving particle is:

$$e = mc^2\sqrt{(1 - v^2/c^2)}$$

This differs slightly from relativity. While relativity distinguishes between the rest energy $m'c^2$ and the energy of a moving particle, Unified Field Theory asserts that the energy when moving at velocity v is still equal to the rest energy $m'c^2$.

Unified Field Theory emphasizes that energy is relative to a specific observer. An observer in the s' frame sees point o at rest with energy $m'c^2$, while an observer in the s frame, seeing point o moving at velocity v, perceives its energy as $mc^2\sqrt{(1 - v^2/c^2)}$. However, no observer can perceive the energy of point o as mc^2.

This view suggests that while different observers see different forms of energy, the total energy remains invariant, which is considered a more reasonable explanation than the relativistic viewpoint.

3. Relationship Between the Unified Field Theory Energy Equation and the Classical Mechanics Kinetic Energy Formula

In classical mechanics, the kinetic energy of a particle with mass m moving at velocity v relative to an observer is given by $E_k = 1/2\ mv^2$. Unified Field Theory shares a similar equation with relativity:

$$(m - m') \ c^2 = Ek$$

Expanding the square root in the Unified Field Theory energy equation $e = mc^2 \sqrt{(1 - v^2/c^2)}$ as a series:

$$\sqrt{(1 - v^2/c^2)} \approx 1 - v^2/2c^2 \cdots$$

and neglecting higher-order terms, we get:

$$e \approx mc^2 - mv^2/2$$

Here, $mv^2/2$ is the kinetic energy Ek from Newtonian mechanics. From $e = m'c^2$, it follows that:

$$mv^2/2 \approx mc^2 - m'c^2 = c^2 \ (m - m')$$

This indicates that classical kinetic energy is the change in rest mass due to motion.

4. Relationship Between Momentum and Kinetic Energy in Unified Field Theory

In Unified Field Theory, the rest momentum is $P'=m'C$, and the momentum during motion is $P=m(C-V)$. The scalar form of momentum is:

$$p = mc\sqrt{\ (1 - v^2/c^2)}$$

Unified Field Theory posits that the magnitude of the rest momentum and the momentum during motion are equal:

$$p = mc\sqrt{\ (1 - v^2/c^2)} \ = m'c$$

where m' is the rest mass, and m is the mass when moving at velocity v.

Unified Field Theory asserts that the total energy of a particle at rest, m'c ^{2}, and during motion, mc^2 — Ek , are equivalent:

$$mc^2 - Ek = m'c^2$$

where Ek $\approx$ (1/2) m v^2 is the kinetic energy.

To explore the relationship between the kinetic energy Ek and momentum P of a photon, we use p^2= m'^{2}c^2 in the equation mc^2— Ek = m'c ^{2}:

$$mc^2 - Ek = p^2/m'$$

For a photon with m'=0, the rest energy m'c ^{2}=0, and the photon's kinetic energy becomes Ek= mc^2.

Dividing the energy equation by the speed of light c, we obtain the photon's momentum equation:

$$P = e/c$$

This shows the relationship between the energy e and momentum P of a photon.

5. Squaring the Unified Field Theory Energy Equation

By squaring both sides of the Unified Field Theory energy equation e = m'c^2 = mc^2$\sqrt{}$ (1 - v^2/c^2) , we get:

$$e^2= m'^2c^2c^2 = m^2c^2c^2 - m^2c^2v^2$$

Rearranging:

$$m^2c^2c^2 = m^2c^2v^2 + m'^2c^2c^2$$

which gives us:

$$m^2c^2c^2 = p'^2c^2 + m'^2c^2c^2$$

This equation looks similar to that in relativity, where $e^2 = m^2c^2c^2$. However, Unified Field Theory sees the total energy differently, as:

$$e^2 = m'^2c^2c^2 = m^2c^2c^2 - p'^2c^2$$

Chapter 40. Photon Model

An accelerating charge, relative to an observer, generates an accelerating and changing electromagnetic field in the surrounding space. This changing electromagnetic field can produce an anti-gravity field, which may cause the mass and charge of the accelerating charge or nearby electrons to disappear. When the mass and charge of an electron vanish, the surrounding force field and electromagnetic characteristics also disappear. This excitation propagates outward at the speed of light, forming what we recognize as electromagnetic waves, or light.

There are two photon models:

1. Single Electron Model

In this model, a single excited electron moves away from the observer in a helical path. The center of its rotation follows a straight line, and the velocity along this straight line is the speed of light.

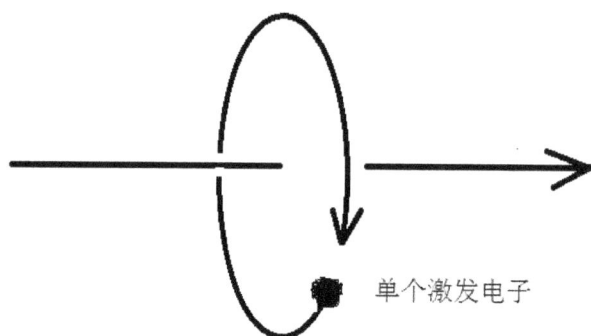

单个激发电子

2. Dual Electron Model

Here, two excited electrons rotate around a straight line while also moving parallel to the line at the speed of light. The resulting motion is a cylindrical helical path that moves away from the observer. The two electrons are symmetric about the center of the straight line, maintaining perpendicular alignment.

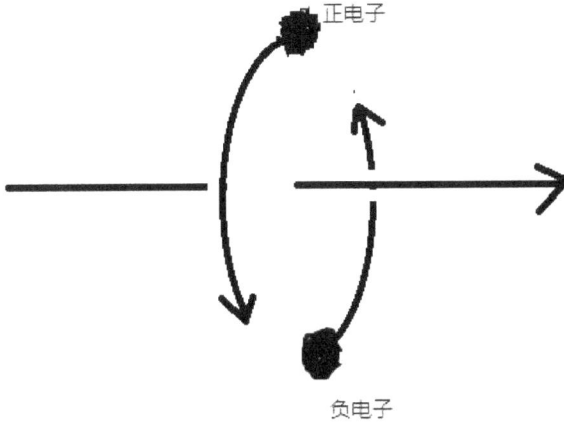

正电子

负电子

The momentum of a photon is given by the equation:

$$\vec{p} = m\,\vec{c}$$

where m is the photon's relativistic mass, and $\vec{c}$ is the vector of the speed of light. Both the rest momentum and rest mass of a photon are zero. The energy of a photon is expressed as:

$$E = m\,c^2$$

When an electron is subjected to the *mass increase force*, $(\vec{c} - \vec{v})\,\frac{dm}{dt}$, it enters an excited state with zero rest mass, which characterizes the photon. A photon always moves at the speed of light relative to any observer.

In the universe, space around any particle radiates outward from the particle at the speed of light. The photon remains stationary within this expanding space and moves along with it. The particle-like nature of the photon arises because it consists of an excited electron, while its wave-like behavior is due to the inherent oscillation of space itself, with the speed of oscillation being the speed of light.

Conclusion

Zhang XiangQian's Unified Field Theory offers a groundbreaking reinterpretation of fundamental physics, challenging the existing frameworks and proposing a new understanding of the universe. By introducing the idea that space itself is in constant motion — moving outward from objects at the speed of light in a cylindrical spiral — this theory redefines the relationships between mass, charge, energy, and the forces of nature. What we perceive as the physical world is revealed to be an illusion shaped by human observation, while the true nature of reality lies in the interactions between objects and space.

A key insight in Zhang's theory addresses a critical gap in current physics: while rest energy is widely recognized, the root cause of rest energy has not been fully understood. Zhang proposes that rest energy is a direct result of rest momentum, stemming from the constant outward movement of space from an object at the speed of light. This revelation reshapes our understanding of energy, mass, and motion, offering new pathways to comprehend the fundamental behavior of matter in the universe.

By uniting the four fundamental forces — electric force, magnetic force, gravitational force, and nuclear force — into a single framework, Zhang's work opens new possibilities for scientific exploration. His insight that, **as mass approaches zero, the speed of an object instantaneously reaches the speed of light and the distance it traverses becomes zero**, introduces profound implications for our understanding of the universe.

As with any paradigm-shifting theory, the road to acceptance may be met with skepticism. However, the potential applications of Zhang's work are too significant to be ignored. His Unified Field Theory not only challenges long-held scientific beliefs but also offers practical avenues for the advancement of human knowledge and civilization.

The Academic Edition of this book, with refined explanations and additional clarity, aims to make these profound concepts accessible to a broader audience. It serves not only as a detailed theoretical treatise but also as a call to explore new horizons in science and technology. The unified vision of space, matter, and motion presented here has the potential to reshape our understanding of the universe and lay the groundwork for future discoveries.

Appendix

1. Main Application of Unified Field Theory — Artificial Field Scanning Technology

Artificial field scanning technology uses positive and negative gravitational fields generated by changing electromagnetic fields, distinct from antigravity as gravity and gravitational fields operate in different dimensions. This technology is controlled by computer programs and serves as a fundamental power source, similar to electrical devices on Earth.

The principle behind artificial field scanning is comparable to Faraday's concept of converting electricity to magnetism and vice versa, through the mutual conversion of electromagnetic fields and gravitational fields. Artificial fields represent an advanced form of electricity that could eventually replace the electrical energy currently in use.

The theoretical foundation for artificial field scanning is grounded in *Unified Field Theory*. For further details, you can reach Zhang Xiangqian via WeChat.

2. Components of Artificial Field Scanning Equipment

Artificial field scanning equipment consists of two core components:

- **Hardware**: The hardware generates and emits artificially created fields that can be directed remotely, even from the sky, towards the ground. These fields can penetrate walls and interact with objects without physical contact.

- **Software**: The software controls the equipment and its applications. Through specific programming, the artificial field can manipulate space, time, and matter.

Artificial field scanning works similarly to a generator: it doesn't create energy but converts other forms (such as electrical or solar energy) into field energy. The transmitted energy can remotely alter an object's mass, velocity, position, temperature, and even the space and time it occupies.

This ability to transmit energy without wires—through a vacuum—is a major advantage of artificial field scanning, offering the potential for centralizing devices and reducing the global demand for physical products. For example, one massive computer could be shared by billions of people in the future.

3. Applications of Artificial Field Scanning

Artificial field scanning has the potential to offer far more than conventional electrical energy. It has the ability to influence spacetime, creating profound technological applications:

1) Light-Speed Travel

By reducing the mass of a spacecraft to zero, artificial field scanning can enable light-speed travel, instantly propelling the spacecraft at the speed of light.

2) Cold Welding in Manufacturing

Artificial field scanning can place materials into a quasi-excited state, allowing them to merge without resistance. Once the field is removed, the objects are welded together, significantly speeding up manufacturing processes and lowering costs.

3) *Non-Invasive Medical Procedures*

Through precise control of fields, artificial scanning can manipulate molecules and atoms, enabling high-speed detection and repair of internal structures without invasive surgery.

4) *Instantaneous Global Movement*

Artificial field scanning could create a global movement network that transports individuals by irradiating them with fields, causing them to disappear from one location and reappear in another. However, this would be limited to planetary distances.

5) *Wireless Power Transmission*

Similar to electricity, artificial fields can transmit power over long distances without physical contact, providing a decentralized energy source for various devices.

6) *Solar Energy Concentration*

By compressing space, artificial fields can concentrate solar energy from a large area, potentially solving the global energy crisis.

7) *Infinite Information Storage*

Artificial field scanning can compress space infinitely, allowing for the storage and transmission of vast amounts of information, revolutionizing data storage technologies.

8) *Virtual Construction*

Artificial fields could create virtual objects and environments, allowing billions of people to share a single virtual device.

9) *Spacetime Refrigeration*

A spacetime refrigerator could preserve food for extended periods by altering the flow of time within the storage unit, surpassing traditional refrigeration methods.

10)   *Consciousness Storage and Transfer*

Artificial field scanning could read and store human consciousness, potentially allowing for the preservation of human memory and knowledge, or even the possibility of eternal life.

4. Steps to Build an Artificial Field Scanning System

1) *Theoretical Foundations*

Define the core equations for electromagnetic and gravitational fields. These equations have already been established.

2) *Field Interaction Equations*

Develop mathematical equations that describe how changing gravitational fields generate electromagnetic fields and vice versa. This step has also been completed.

3) *Experiment Design*

Conduct experiments to verify that changing electromagnetic fields can generate positive and negative gravitational fields. Significant progress has already been made. In 2023, it was discovered that accelerating charges can generate a gravitational field in the opposite direction of acceleration, while in 2024, it was found that changing magnetic fields can generate vortex gravitational fields.

4) *Refinement of Equations*

Based on experimental results, refine the application equations, especially those quantifying how charge and acceleration produce gravitational fields. This will guide the construction of artificial field scanning devices.

5) *Software Development*

Design computer programs to control the artificial field scanning devices. Different applications require specialized software, from moving objects to medical procedures, with each function controlled by the software.

6) *Widespread Application*

Expand the use of artificial field scanning to replace electrical energy and integrate it into other fields, such as space exploration and construction.

5. Conclusion

Artificial field scanning is a groundbreaking scientific endeavor with the potential to revolutionize human technology. The development process could be comparable in scale to the Manhattan Project, yet the key discovery—proving that changing electromagnetic fields generate gravitational fields—has already been achieved. Artificial fields operate at normal temperatures, making them practical for a wide range of applications.

The development of this technology will require collaboration from universities and research institutions. With concerted effort, many of the key applications for artificial field scanning could be realized within 1 to 5 years.

About the Author

Zhang XiangQian was born in 1967 in Lujiang County, Anhui, China. With a background shaped by modest beginnings and limited formal education, Zhang's life took a remarkable turn in the summer of 1985 when he claims to have encountered an extraterrestrial civilization. Despite having only a middle school education and working as a farmer, Zhang returned from this extraordinary experience with profound insights into the fundamental truths of the universe. His knowledge spans a range of complex subjects, including time, space, mass, charge, fields, the speed of light, momentum, energy, force, and motion.

Building on these revelations, Zhang formulated what he calls the **Grand Unified Equation of the Universe**, a theoretical framework that seeks to unify the four fundamental forces of nature — **electric force, magnetic force, gravitational force, and nuclear force** — into a single coherent expression. In addition to this groundbreaking discovery, Zhang has uncovered what he refers to as the core secrets of the universe, which encompass the **Unified Field Theory**, cosmic spatial information fields, the technology behind light-speed UFOs, and artificial field scanning methods.

Currently residing in Erlong New Street, Tongda Town, Lujiang County, Zhang works as a welder and bicycle repairman. In his spare time, he is dedicated to advancing and sharing his theories on the Unified Field and artificial field scanning technology. Despite leading a modest lifestyle, his work has the potential to challenge and transform our understanding of the universe.

Contact   **Information**:

Please note that Zhang XiangQian does not speak or write in English. His translator and editor, Lynn Beran, is available for ongoing contact and inquiries.

WeChat (available to China-based accounts): 18714815159 Email: zzqq2100@163.com (please use Chinese for communication) For English inquiries, please contact Lynn Beran at newlifenewhope@gmail.com.

References

1. Zhang Sanhui, *University Physics (Third Edition) Mechanics and Electromagnetism*, Tsinghua University Press, 2009/2.

www.ingramcontent.com/pod-product-compliance
Lightning Source LLC
Chambersburg PA
CBHW081814200326
41597CB00023B/4246